Medical biotechnology: Achievements, prospects and perceptions

Albert Sasson

United Nations University Press

TOKYO · NEW YORK · PARIS

United Nations University Press
United Nations University, 53-70, Jingumae 5-chome,
Shibuya-ku, Tokyo, 150-8925, Japan
Tel: +81-3-3499-281   Fax: +81-3-3406-7345
E-mail: sales@hq.unu.edu
General enquiries: press@hq.unu.edu
http://www.unu.edu

United Nations University Office at the United Nations, New York
2 United Nations Plaza, Room DC2-2062, New York, NY 10017, USA
Tel: +1-212-963-6387   Fax: +1-212-371-9454
E-mail: unuona@ony.unu.edu

United Nations University Press is the publishing division of the United Nations University.

Cover design by Rebecca S. Neimark, Twenty-Six Letters

Printed in Hong Kong

ISBN 92-808-1114-2

Library of Congress Cataloging-in-Publication Data

Sasson, Albert.
 Medical biotechnology : achievements, prospects and perceptions / Albert Sasson.
  p. ; cm.
 Includes bibliographical references and index.
 ISBN 9280811142 (pbk.)
 1. Biotechnology. 2. Biotechnology industries. 3. Pharmaceutical biotechnology. [DNLM: 1. Biotechnology. 2. Technology, Pharmaceutical. QV 778 S252m 2005] I. Title.
 TP248.2.S273   2005
 660.6—dc22   2005018003

Medical biotechnology

This book was produced as an outcome of the United Nations University Institute of Advanced Studies (UNU-IAS) programme on Science Policy for Sustainable Development.

The UNU-IAS was inaugurated in April 1996. Its mission statement is "Advancing knowledge and promoting learning for policymaking to meet the challenges of sustainable development".

UNU-IAS conducts research, postgraduate education, and capacity development, both in-house and in cooperation with an interactive network of academic institutions and international organizations. The Institute's research concentrates on exploring the key catalysts and drivers of sustainable development, which often depend on our capacity to harmonize, if not optimize, the interaction between societal and natural systems. This includes the development and use of new technologies and approaches; the study of major trends and pressures such as urbanization, regionalization, and globalization; as well as the exploration of integrated approaches to policy-making, decision-making, and sustainable development governance.

UNU-IAS website: http://www.ias.unu.edu

UNITED NATIONS UNIVERSITY

UNU-IAS
Institute of Advanced Studies

Contents

Foreword

Since the advent of civilization, biotechnology has gradually become commonplace around the world and it has developed into a global multi-billion dollar industry. But, as with many new technologies, society's ability to manage, share and regulate the development and use of modern biotechnology poses many challenges, opportunities and even risks. The implications of modern biotechnology connote a broad coverage of sectors and issues such as human health, industrial applications, food safety, and the life and health of plants and animals. The risks could include everything from genetically modified organisms, alien species and introduced plant and animal pests, to the erosion of biodiversity, toxic weapons of war, utilization of genetic resources and "mad cow" disease.

The UNU-IAS Programme on Science Policy for Sustainable Development is elaborating the challenges that are facing modern society through the advances in modern biotechnology. It is doing so by exploring the implications, impacts, perceptions and prospects for ensuring the secure and wise use in the future of products derived from modern biotechnology.

This book by Professor Albert Sasson, *Medical Biotechnology: Achievements, Prospects and Perceptions*, is the first in a series of publications that UNU-IAS is undertaking to present these broad implications of modern biotechnology and takes stock of the advances that have been made in recent years.

The book first looks at the drivers of modern biotechnology development in the United States, the European Union and Japan and at what progress has been made in the development of biotechnology to fight major global health concerns such as Ebola fever, the human immuno-deficiency virus, the SARS virus and the Avian flu virus, before turning to some of the regulatory and public perception concerns. The book also provides a state of the art analysis of the progress of selected developing countries in the development of their own biotech industries. Finally, Professor Sasson examines some of the most controversial areas of modern biotechnology, including issues such as stem cell research and gene therapy and some of the ethical issues they raise.

The findings of this book are a valuable contribution to the state of our knowledge about modern biotechnology, to the UNU-IAS efforts to raise awareness among policy makers and stakeholders, and to educating the public at large about the greater implications and prospects concerning the advances of this rapidly growing new technology.

A. H. Zakri
Director, UNU-IAS

1

Introduction: Biotechnology, bio-industry and bio-economy

The word "biotechnology" was coined in 1919 by Karl Ereky, a Hungarian engineer, to refer to methods and techniques that allow the production of substances from raw materials with the aid of living organisms. A standard definition of biotechnology was reached in the Convention on Biological Diversity (1992) – "any technological application that uses biological systems, living organisms or derivatives thereof, to make or modify products and processes for specific use". This definition was agreed by 168 member nations, and also accepted by the Food and Agriculture Organization of the United Nations (FAO) and the World Health Organization (WHO).

Biotechnologies therefore comprise a collection of techniques or processes using living organisms or their units to develop added-value products and services. When applied on industrial and commercial scales, biotechnologies give rise to bio-industries. Conventional biotechnologies include plant and animal breeding and the use of micro-organisms and enzymes in fermentations and the preparation and preservation of products, as well as in the control of pests (e.g. integrated pest control). More advanced biotechnologies mainly relate to the use of recombinant deoxyribonucleic acid (DNA) techniques (i.e. the identification, splicing and transfer of genes from one organism to another), which are now supported by research on genetic information (genomics). This distinction is merely a convenience, because modern techniques are used to improve conventional methods; for example, recombinant enzymes and genetic markers are employed to improve fermentations and plant and animal breeding. It is, however, true that the wide range of biotechnologies, from the simplest to the most sophisticated, allows each country to select those that suit its needs and development priorities, and by doing so even reach a level of excellence (for example, developing countries that have used *in vitro* micro-propagation and plant-tissue cultures to become world-leading exporters of flowers and commodities).

The potential of biotechnology to contribute to increasing agricultural, food and feed production, improving human and animal health, mitigating pollution and protecting the environment was acknowledged in *Agenda 21* – the work programme adopted by the 1992 United Nations Conference on Environment and Development in Rio de Janeiro. In 2001, the *Human Development Report* considered biotechnology to be the means to tackle major health challenges in poor countries, such as infectious diseases (tuberculosis), malaria and HIV/AIDS, and an adequate tool to aid the development of the regions left behind by the "green revolution"; these are home to more than half the world's poorest populations, who depend on agriculture, agroforestry and livestock husbandry. New and more effective vaccines, drugs and diagnostic tools, as well as more food and feed of high nutritional value, will be needed to meet the expanding needs of the world's populations.

Biotechnology and bio-industry are becoming an integral part of the knowledge-based economy, because they are closely associated with progress in the life sciences and in the applied sciences and technologies linked to them. A new model of economic activity is being ushered in – the bio-economy – in which new types of enterprise are created and old industries are revitalized. The bio-economy is defined as including all industries, economic activities and interests organized around living systems. The bio-economy can be divided into two primary industry segments: the bio-resource industries, which directly exploit biotic resources – crop production, horticulture, forestry, livestock and poultry, aquaculture and fisheries; and related industries that have large stakes as either suppliers to or customers of the bio-resource sector – agrochemicals and seeds, biotechnologies and bio-industry, energy, food and fibre processing and retailing, pharmaceuticals and health care, banking and insurance. All these industries are closely associated with the economic impact of human-induced change to biological systems (Graff and Newcomb, 2003).

The potential of this bio-economy to spur economic growth and create wealth by enhancing industrial productivity is unprecedented. It is therefore no surprise that high-income and technologically advanced countries have made huge investments in research and development (R&D) in the life sciences, biotechnology and bio-industry. In 2001, bio-industries were estimated to have generated US$34.8 billion in revenues worldwide and to employ about 190,000 people in publicly traded firms. These are impressive results given that, in 1992, bio-industries were estimated to have generated US$8.1 billion and employed fewer than 100,000 persons.

The main beneficiaries of the current "biotechnology revolution" and the resulting bio-industries are largely the industrialized and technologically advanced countries, i.e. those that enjoy a large investment of their domestic product in R&D and technological innovation. Thus, the United States, Canada and Europe account for about 97 per cent of the global biotechnology revenues, 96 per cent of persons employed in biotechnology ventures and 88 per cent of all biotechnology firms. Ensuring that those who need biotechnology have access to it therefore remains a major challenge. Similarly, creating an environment conducive to the acquisition, adaptation and diffusion of biotechnology in developing countries is another great challenge. However, a number of developing countries are increasingly using biotechnology and have created a successful bio-industry, at the same time increasing their investments in R&D in the life sciences.

According to the Frost & Sullivan Chemicals Group in the United Kingdom, some 4,300 biotechnology companies were active globally in 2003: 1,850 (43 per cent) in North America; 1,875 (43 per cent) in Europe; 380 (9 per cent) in Asia; and 200 (5 per cent) in Australia. These companies cover the gamut from pure R&D participants to integrated manufacturers to contract manufacturing organizations (CMOs). The United States has the largest number of registered biotechnology companies in the world (318), followed by Europe (102). In 2002, the annual turnover of these companies was US$33.0 billion in the United States and only US$12.8 billion in Europe. Some US$20.5 billion was allocated to research in the United States, compared with US$7.6 billion in Europe (Adhikari, 2004).

US biotechnology and bio-industry

The consultancy firm Ernst & Young distinguishes between US companies that produce medicines and the others. The former include pioneers such as Amgen, Inc., Genentech, Inc., Genzyme Corporation, Chiron Corporation and Biogen, Inc. The annual turnover of these five companies represents one-third of the sector's total (US$11.6 billion out of US$33.0 billion); in addition, their product portfolio enables them to compete with the big pharmaceutical groups in terms of turnover and stock value. For instance, Amgen, with US$75 billion market capitalization, is more important than Eli Lilly & Co., and Genentech's market capitalization is twice that of Bayer AG (Mamou, 2004e).

In 2002, Amgen, had six products on the market producing global revenues of US$4,991 million. Genentech was in second place with 11 products on the market and revenues worth US$2,164 million. The remaining places in the top five were filled by Serono SA (six products, US$1,423 million), Biogen (two products, US$1,034 million) and Genzyme Corporation (five products, US$858 million) (Adhikari, 2004).

Over the past decade, a clutch of companies has amassed significant profits from a relatively limited portfolio of drugs. There is, today, heightened recognition that lucrative opportunities await companies that can develop even a single life-saving biotechnology drug. For instance, Amgen's revenues increased by over 40 per cent between 2001 and 2002 owing to the US$2 billion it made in 2002 from sales of Epogen and the US$1.5 billion earned from sales of Neupogen. Over US$1 billion in sales of Rituxan – a monoclonal antibody against cancer – in 2002 helped Genentech record a 25 per cent growth over its 2001 performance (Adhikari, 2004).

In California, there are two biotechnology "clusters" of global importance: one in San Diego–La Jolla, south of Los Angeles, and the other in the Bay Area, near San Francisco. A cluster is defined as a group of enterprises and institutions in a particular sector of knowledge that are geographically close to each other and networked through all kinds of links, starting with those concerning clients and suppliers. In neither biotechnology cluster does it take more than 10 minutes to travel from one company to another. The San Diego cluster is supported in all aspects of its functioning, including lobbying politicians and the various actors in the bio-economy, by Biocom – a powerful association of 450 enterprises, including about 400 in biotechnology, in the San Diego region. The cluster relies on the density and frequency of exchanges between industry managers and university research centres. For instance, one of its objectives is to shorten the average time needed to set up a licensing contract between a university and a biotechnology company; it generally takes 10 months to establish such a contract, which is considered too long, so the cluster association is bringing together all the stakeholders to discuss this matter and come to a rapid conclusion (Mamou, 2004e).

The clusters have developed the proof of concept, to show that from an idea, a theory or a concept there could emerge a business model and eventually a blockbuster drug. Such an endeavour between the researchers and bio-industry would lead to licensing agreements that rewarded the discovery work. A strategic alliance between politics, basic research and the pharmaceutical industry (whether biotechnological or not) within the cluster would be meaningless without capital. In fact, bio-industries' success is above all associated with an efficient capital market, according to David Pyott, chief executive officer of Allergan, the world leader in ophthalmic products and the unique owner of Botox – a product used in cosmetic surgery and the main source of the company's wealth. No cluster can exist without a dense network of investors, business angels, venture capitalists and bankers, ready to get involved in the setting up of companies (Mamou, 2004e).

The two Californian clusters represented 25.6 per cent of US companies in 2001. The corresponding figures for other states were as follows: Massachusetts, 8.6 per cent; Maryland, 7.7 per cent; New Jersey, 5.9 per cent; North Carolina, 5.8 per cent; Pennsylvania, 4.6 per cent; Texas, 3.4 per cent; Washington, 3.1 per cent; New York, 3.1 per cent; Wisconsin, 2.5 per cent; the rest of the country accounted for the remaining 29.7 per cent (data from the US Department of Commerce Technology Administration and Bureau of Industry and Security).

Europe's biotechnology and bio-industry

The European bio-industry is less mature than its US counterpart. Actelion of Switzerland qualified as the world's fastest-growing drugs group in sales terms following the launch of its first drug, Tracleer, but it did not achieve profitability until 2003. Similarly, hardly any European biotechnology companies are earning money. Only Serono SA – the Swiss powerhouse of European biotechnology – has a market capitalization to rival US leaders (Firn, 2003). Serono SA grew out of a hormone extraction business with a 50-year record of profitability and is the world leader in the treatment of infertility; it is also well known in endocrinology and the treatment of multiple sclerosis. In 2002, Serono SA made US$333 million net profit from US$1,546 million of sales; 23 per cent of the revenue from these sales was devoted to its R&D division, which employs 1,200 people. The Spanish subsidiary of Serono SA in Madrid is now producing recombinant human growth hormone for the whole world, whereas factories in the United States and Switzerland have ceased to produce it. The Spanish subsidiary had to invest €36 million in order to increase its production, as well as another €5 million to upgrade its installations for the production of other recombinant pharmaceuticals to be exported worldwide.

In spite of a wealth of world-class science, the picture in much of Europe is of an industry that lacks the scale to compete and is facing the financial crunch, which may force many companies to seek mergers with stronger rivals (Firn, 2003).

Germany

Germany has overtaken the United Kingdom and France, and is currently home to more biotechnology companies than any country except the United States. But, far from pushing the boundaries of biomedical science, many companies are putting cutting-edge research on hold and are selling valuable technology just to stay solvent. Until the mid-1990s, legislation on genetic engineering in effect ruled out the building of a German bio-industry. According to Ernst & Young, the more than 400 companies set up in Germany since then needed to raise at least US$496 million from venture capitalists over 2004 to refinance their hunt for new medicines. Most were far from having profitable products and, with stock markets in effect closed to biotechnology companies following the bursting of the bubble in 2000, they were left to seek fourth or even fifth rounds of private financing (Firn, 2003).

The biggest German biotechnology companies, such as GPC Biotech and Medigene, were able to raise significant sums in initial public offerings at the peak of the Neuer Markt, Germany's market for growth stocks. But when the technology bubble burst in 2000, it became clear to GPC Biotech that investors put very little value on "blue-sky" research. "They wanted to see proven drug candidates in clinical trials", said Mirko Scherer, chief financial officer (cited in Firn, 2003). The only option for companies such as GPC Biotech and Medigene was to buy drugs that could be brought to market more quickly. GPC Biotech has used the cash it earned from setting up a research centre for Altana, the German chemicals and pharmaceutical group, to acquire the rights to satraplatin, a cancer treatment that was in the late stages of development. In October 2003, regulators authorized the initiation of the final round of clinical trials (Firn, 2003). After a series of clinical setbacks, Medigene has mothballed its early-stage research to cut costs and has licensed in late-stage products to make up for two of its own drugs that failed. The strategy will help the company eke out its cash; but cutting back on research will leave little in its pipeline (Firn, 2003).

Many of Germany's biotechnology companies have abandoned ambitious plans to develop their own products and chosen instead to license their drug leads to big pharmaceutical companies in exchange for funding that will allow them to continue their research. This approach is supported by the acute shortage of potential new medicines in development by the world's biggest pharmaceutical companies. But Germany's bio-industry has few experimental drugs to sell – about 15 compared with the more than 150 in the United Kingdom's more established industry. Moreover, most of Germany's experimental drugs are in the early stages of development, when the probability of failure is as high as 90 per cent. That reduces the price that pharmaceutical companies are willing to pay for them (Firn, 2003).

Companies also have to struggle with less flexible corporate rules than their rivals in the United Kingdom and the United States. Listed companies complain that the Frankfurt stock exchange does not allow injections of private equity, which are common in US biotechnology. As a result, few of Germany's private companies state that they expect to float in Frankfurt. Most are looking to the United States, the United Kingdom or Switzerland, where investors are more comfortable with high-risk stocks. However, many German companies may not survive long enough to make the choice (Firn, 2003).

Faced with this bleak outlook, many in the industry agree that the only solution is a wave of consolidation that will result in fewer, larger companies with more diverse development pipelines. A number of investors in Germany's bio-industry are already pushing in this direction. TVM, the leading German venture capital group, had stakes in 14 German biotechnology companies and was trying to merge most of them. TVM sold off all Cardion's drug leads after failing to find a merger partner for the arthritis and transplant medicine specialists. After raising US$14.1 million in 2002, Cardion has become a shell company that may one day earn royalties if its discoveries make it to market. UK-based Apax Partners was said to have put almost its entire German portfolio up for sale. The fate of MetaGene Pharmaceuticals, one of Apax's companies, may await many others. In October 2003, the company was bought by the British Astex, which planned to close the German operation after stripping out its best science and its US$15 million bank balance (Firn, 2003).

GPS Biotech's chief financial officer was critical of the investors who turned their backs on Germany and put 90 per cent of their funds in the United States, when a lot of European companies were very cheap. And although Stephan Weselau, chief financial officer of Xantos, was frustrated that venture capitalists saw little value in his young company's anti-cancer technology, he was adamant about the need for Germany's emerging biotechnology to consolidate if it was to compete against established companies in Boston and San Diego (Firn, 2003).

The United Kingdom

The market for initial public offerings in the United Kingdom was all but closed to biotechnology for the three-year period 2000–2002; it reopened in the United States in 2003. City of London institutions, many of which took huge losses on biotechnology, were reluctant to back new issues and have become more fussy about which quoted companies they are prepared to finance (Firn, 2003).

The United Kingdom is home to one-third of Europe's 1,500 biotechnology companies and more than 40 per cent of its products in development. Although the United Kingdom had 38 marketed biotechnology products and 7 more medicines awaiting approval by the end of 2003, analysts stated that there were too few genuine blockbusters with the sort of sales potential needed to attract investors' attention away from the United States. A dramatic case is that of PPL (Pharmaceutical Proteins Ltd) Therapeutics – the company set up to produce drugs in the milk of a genetically engineered sheep (Polly). By mid-December 2003, the company had raised a paltry US$295,000 when auctioneers put a mixed catalogue of redundant farm machinery and laboratory equipment under the hammer. This proved that exciting research (Dolly and Polly sheep) does not always lead to commercial success (Firn, 2003).

The profitable British companies reported pre-tax profits of £145 million in 2003, less than 15 per cent of the US$1.9 billion pre-tax profits reported by Amgen. By mid-2003, the British biotechnology sector seemed to be coming of age. Investors could choose between three companies that had successfully launched several products and boasted market capitalizations in excess of US$884 million. Since then they have seen PowderJect Pharmaceuticals plc acquired by Chiron Corp., the US vaccines group, for a deal value of £542 million in May 2003; and General Electric swooped in with a £5.7 billion bid for Amersham, the diagnostics and biotechnology company, in October 2003. Earlier, in July 2000, Oxford Asymmetry had been purchased by the German company Evotec Biosystems for £343 million, and, in September 2002, Rosemont Pharma was acquired by the US firm Bio-Technology General for £64 million (Dyer, 2004).

In May 2004, Union Chimique Belge (UCB) agreed to buy Celltech, the United Kingdom's biggest biotechnology company, for £1.53 billion (€2.26 billion). UCB decided Celltech could be its stepping stone into biotechnology after entering an auction for the marketing rights to Celltech's new treatment for rheumatoid arthritis (CPD 870), touted as a blockbuster drug with forecast annual sales of more than US$1 billion. After seeing trial data not revealed to the wider market, UCB decided to buy the whole company. The surprise acquisition was accompanied by a licensing deal that gives UCB the rights to CPD 870, which accounted for about half the company's valuation. Göran Ando, the Celltech chief executive who will become deputy chief executive of UCB, stated: "we will immediately have the financial wherewithal, the global commercial reach and the R&D strength to take all our drugs to market." News of the deal, which will be funded with debt, sent Celltech shares 26 per cent higher to £5.42, whereas UCB shares fell 4 per cent to €33.68 (Firn and Minder, 2004).

Celltech had been the grandfather of the British biotechnology sector since it was founded in 1980. With a mixture of seed funding from the Thatcher government and the private sector, the company was set up to commercialize the discovery of monoclonal antibodies that can become powerful medicines. Listed in 1993, the company made steady progress in its own research operations, but gained products and financial stability only with the acquisitions of Chiroscience in 1999 and Medeva in 2000. It also acquired Oxford GlycoSciences in May 2003 in a deal worth £140 million. The great hopes Celltech has generated were based largely on CPD 870, the arthritis drug it planned to bring to market in 2007 that could be by far the best-selling product to come out of a British biotechnology company. After the UCB–Celltech deal, the group ranked fifth among the top five biopharmaceutical companies, behind Amgen, €6.6 billion in revenue in 2003; Novo Nordisk, €3.6 billion; Schering, €3.5 billion; and Genentech, €2.6 billion (Dyer, 2004; Firn and Minder, 2004). Based on 2003 results, the combined market capitalization of UCB Pharma and Celltech will be €7.14 billion; revenues, €2,121 million; earnings before interest, tax and amortization, €472 million; pharmaceutical R&D budget, €397 million; number of employees, approximately 1,450 (Firn and Minder, 2004).

Celltech is the biggest acquisition by UCB, which branched out from heavy chemicals only in the 1980s. Georges Jacob, its chief executive since 1987, stated that when he joined UCB he found a company "devoted to chemicals, dominated by engineers, pretty old-fashioned and very much part of heavy industry". UCB had been built entirely on internal growth, and its only other sizeable acquisition was the speciality chemicals business of US-based Solutia in December 2002 for US$500 million, a move that split the Belgian group's €3 billion revenues evenly between pharmaceuticals and chemicals. One constant was the continued presence of a powerful family shareholder, owning 40 per cent of UCB's equity via a complicated holding structure (Firn and Minder, 2004).

UCB made its first foray into pharmaceuticals in the 1950s with the development of a molecule it sold to Pfizer, Inc. This became Atarax, an anti-histamine used to relieve anxiety. The relationship with Pfizer was revived in a more lucrative fashion for UCB following the 1987 launch of Zyrtec, a blockbuster allergy treatment that Pfizer helped to distribute in the United States. Although UCB has a follow-up drug to Zyrtec, it faces the loss of the US patent in 2007. UCB also had to fight patent challenges to its other main drug, Keppra, an epilepsy treatment. With the takeover of Celltech, UCB will gain a pipeline of antibody treatments for cancer and inflammatory diseases to add to its allergy and epilepsy medicines. According to most analysts, the expansion in health-care activities will lead the group to divest itself of its remaining chemical business (Firn and Minder, 2004).

After this takeover and following the earlier acquisition of PowderJect Pharmaceuticals and Amersham by US companies, there is not much left in the United Kingdom's biotechnology sector except Acambis, another vaccine-maker, valued at about £325 million, and a string of companies below the £200 million mark where liquidity can be a problem for investors. The industry was therefore afraid it would be swamped by its much larger rivals. Martyn Postle, director of Cambridge Healthcare and Biotech, a consultancy, stated that "we could end up with the UK performing the role of the research division of US multinationals" (cited in Dyer, 2004). According to the head of the Bioindustry Association (BIA), "it is clearly the fact that US companies are able to raise much, much more money than in the United Kingdom, which puts them in a much stronger position" (cited in Dyer, 2004). The BIA called for changes in the rules on "pre-emption rights", which give existing shareholders priority in secondary equity offerings. Because Celltech was by far the most liquid stock in the sector, there could be a broader impact on the way the financial sector treats biotechnology, including a reduction in the number of specialist investors and analysts covering the sector (Dyer, 2004).

It is important for the United Kingdom to create an environment in which biotechnology can flourish. The industry has called for institutional reform, including measures to make it easier for companies to raise new capital. The British government must also ensure that its higher education system continues to produce world-class scientists. That reinforces the need for reforms to boost the funding of universities. The Celltech takeover need not be seen as a national defeat for the United Kingdom. The combined company may end up being listed in London. Even if it does not, Celltech's research base in the United Kingdom will expand. Its investors have been rewarded for their faith and, if its CPD 870 drug is approved, UCB's shareholders will also benefit. But, for Celltech's executives, the acquisition is a victory for Europe. The takeover creates an innovative European biotechnology company that is big enough, and has sufficient financial resources, to compete globally. "The key was to have viable European businesses that have a sustainable long-term presence," stated Göran Ando, who confirmed that UCB's research will be run from Celltech's old base in Slough (cited in Dyer, 2004). A lot of hopes are riding on the success of UCB and Celltech, which would allow the fledgling bio-industry to thrive in Europe and prevent the life sciences from migrating to the United States (Dyer, 2004).

France

In France in 2003, according to the France Biotech association, there were 270 biotechnology companies focused on the life sciences and less than 25 years old. They employed 4,500 people – a number that could be multiplied four or five times if about €3 billion were to be invested in public research over three years. In 2003, France invested only €300 million of private funds and €100 million of public funds in biotechnology, far behind Germany and the United Kingdom, which each invested about €900 million per year. In 2003, France launched a five-year Biotech Plan aimed at restoring the visibility and attractiveness of France in 2008–2010. Three areas – human health, agrifood and the environment – were expected to attract the funds as well as the efforts of universities, public and private laboratories, hospitals, enterprises and investors (Kahn, 2003b).

SangStat, a biotechnology company created in 1989 in the Silicon Valley by Philippe Pouletty (a French medical immunologist), is working on organ transplants. It was established in California because, at the time of its creation, venture capital in France was only just starting to support such endeavours in biotechnology. Between FFr 600 million and FFr 2 billion were needed to set up a biotechnology corporation to develop one or perhaps two new drugs, and bankruptcy was very likely in France. SangStat is now a world leader in the treatment of the rejection of organ transplants and intends to extend its expertise and know-how to the whole area of transplantation. It is already marketing two drugs in the United States and three in Europe (Lorelle, 1999a).

A second corporation, DrugAbuse Sciences (DAS), was established by Pouletty in 1994, by which time venture capital was becoming a more common practice in Europe. Two companies were created at the same time: DAS France and DAS US in San Francisco, both belonging to the same group and having the same shareholders. Being established in Europe and the United States, greater flexibility could be achieved from the financial viewpoint and better resilience to stock exchange fluctuations. DAS was able to increase its capital by FFr 140 million (€21.3 million) in 1999 with the help of European investors (Lorelle, 1999a).

DAS specializes in drug abuse and alcoholism. Its original approach was to study neurological disorders in the patient so as to promote abstinence, treat overdoses and prevent dependence through new therapies. Pouletty had surveyed 1,300 existing biotechnology companies in 1994 and found that hundreds were working on cancer and dozens on gene therapy, diabetes, etc., but not one was working on drug and alcohol addiction. Even the big pharmaceutical groups had no significant activity in this area, although drug and alcohol addiction is considered the greatest problem for public health in industrialized countries. For instance, 2.5 per cent of the annual gross domestic product in France is spent on these illnesses, and some US$250 billion in the United States (Lorelle, 1999a).

A first product, Naltrel, improves on the current treatment of alcoholism by naltrexone. The latter, to be efficient, must be taken as pills every day. But few alcoholics can strictly follow this kind of treatment. In order to free patients from this daily constraint, a monthly intramuscular injection of a delayed-action micro-encapsulated product has been developed, which helps alcoholics and drug addicts to abstain from their drug. The molecule developed inhibits the receptors in the brain that are stimulated by opium-related substances.

Another successful product, COC-AB, has been developed for the emergency treatment of cocaine overdoses. This molecule recognizes cocaine in the bloodstream and traps it before it reaches the brain; it is then excreted through the kidneys in urine. Commercialization of the medicine was expected to help the 250,000 cocaine addicts who are admitted annually to the medical emergency services. In the long term, DAS intends to develop preventive compounds that can inhibit the penetration of the drug into the brain (Lorelle, 1999a).

DAS was expected to become a world-leading pharmaceutical company by 2005–2007 in the treatment of alcoholism and drug addiction or abuse. This forecast was based on the current figures of 30 million chronic patients in the United States and Europe, comprising 22 million alcoholics, 6 million cocaine addicts and 2 million heroin addicts (Lorelle, 1999a).

Another success story is the French biotechnology company Eurofins, founded in Nantes in 1998 to exploit a patent filed by two researchers from the local faculty of sciences. Eurofins currently employs 2,000 people worldwide and in four years increased its annual turnover 10-fold (to €162 million). Its portfolio contains more than 5,000 methods of analysing biological substances. The company is located in Nantes, where 130 people carry out research on the purity and origin of foodstuffs. Despite the closure of some of Eurofins' 50 laboratories in order to improve the company's financial position in the face of the slowdown in the economy, Eurofins wants to continue to grow.

This success story has led the city of Nantes to think about creating a biotechnology city. It has also given a strong impetus to medical biotechnology at Nantes' hospital, where the number of biotechnology researchers soared from 70 to 675. In October 2003, the Institute of Genetics Nantes Atlantique initiated the analysis of human DNA for forensic purposes. This institute, which received venture capital from two main sources, was expected to employ 50 people within two years in order to meet the demand generated by the extension of the national automated database of genetic fingerprinting (Luneau, 2003).

Spain

Oryzon Genomics is a genomics company based in Madrid. It applies genomics to new cereal crops, grapevines and vegetables, as well as to the production of new drugs (especially for Parkinson's and Alzheimer's diseases). It is a young enterprise, an offshoot of the University of Barcelona and the Spanish Council for Scientific Research (CSIC), located in Barcelona's Science Park. With a staff of 22 scientists, the company is experiencing rapid growth and is developing an ambitious programme of functional genomics. It was the first genomics enterprise to have access to special funding from the NEOTEC Programme, in addition to financial support from the Ministry of Science and the Generalitat of Catalonia. Moreover, the National Innovation Enterprise (ENISA), which is part of the General Policy Directorate for Medium and Small Sized Enterprises of the Ministry of the Economy, has invested €400,000 in Oryzon Genomics – this was ENISA's first investment in the biotechnology sector. At the end of 2002, Najeti Capital, a venture capital firm specializing in investments in technology, acquired 28 per cent of Oryzon Genomics in order to support the young corporation. In 2003, Oryzon Genomics' turnover was estimated at €500,000, and its clients comprised several agrifood and pharmaceutical companies as well as public research centres.

Japan's biotechnology and bio-industry

Japan is well advanced in plant genetics and has made breakthroughs in rice genomics, but it is lagging behind the United States in human genetics. Its contribution to the sequencing of the human genome (by teams of researchers from the Physics and Chemistry Research Institute of the Science and Technology Agency, as well as from Keio University Medical Department) was about 7 per cent. In order to reduce the gap with the United States, the Japanese government has invested significant funds in the Millennium Project, launched in April 2000. The project covers three areas: the rice genome, the human genome and regenerative medicine. The 2000 budget included ¥347 billion for the life sciences. The genomics budget, amounting to ¥64 billion, was twice that of the neurosciences. Within the framework of the Millennium Project, the Ministry of Health aimed to promote the study of genes linked with such diseases as cancer, dementia, diabetes and hypertension; results for each of these diseases were expected by 2004 (Pons, 2000).

The Ministry of International Trade and Industry (MITI) set up a Centre for Analysis of Information Relating to Biological Resources. This had a very strong DNA-sequencing capacity – equivalent to that of Washington University in the United States (sequencing of over 30 million nucleotide pairs per annum) – and will analyse the genome of micro-organisms used in fermentation and provide this information to the industrial sector. In addition, following the project launched in 1999 by Hitachi Ltd, Takeda Chemical Industries and Jutendo Medical Faculty aimed at identifying the genetic polymorphisms associated with allergic diseases, a similar project devoted to single-nucleotide polymorphisms (SNPs) was initiated in April 2000 under the aegis of Tokyo University and the Japanese Foundation for Science. The research work is being carried out in a DNA-sequencing centre to which 16 private companies send researchers with a view to contributing to the development of medicines tailored to individuals' genetic make-up. This work is similar to that undertaken by a US–European consortium (Pons, 2000).

On 30 October 2000, the pharmaceutical group Daiichi Pharmaceutical and the giant electronics company Fujitsu announced an alliance in genomics. Daiichi and Celestar Lexico Science (Fujitsu's biotechnology division) were pooling their research efforts over the five-year period 2000–2005 to study the genes involved in cancer, ageing, infectious diseases and hypertension. Daiichi devoted about US$100 million to this research in 2001–2002, and about 60 scientists were involved in this work of functional genomics (Pons, 2000).

On 31 January 2003, the Japan Bioindustry Association (JBA) announced that, as of December 2002, the number of "bioventures" in Japan totalled 334 firms. This announcement was based on a survey – the first of its kind – conducted by the JBA in 2002 to have a better understanding of the nation's bio-industry. A "bioventure" was defined as a firm that employs, or develops for, biotechnology applications; that complies with the definition of a small or medium-sized business as prescribed by Japanese law; that was created 20 years ago; and that does not deal primarily in sales or imports/exports. The 334 bioventures had a total of 6,757 employees (including 2,871 R&D staff), sales amounting to ¥105 billion and R&D costs estimated at ¥51 billion (Japan Bioindustry Association, 2003). The average figures per bioventure were: 20 employees (including 8.6 R&D staff), sales worth ¥314 million and R&D costs of ¥153 million.

The three regions with the highest concentrations of bioventures were Kanto (191, or 57 per cent of the national total), Kinki/Kansai (55, or 16 per cent) and Hokkaido (32, or 10 per cent). One-third of all ventures (112) were located in Tokyo (within the Kanto region). The most common field of bioventure operations was pharmaceuticals and diagnostic product development (94 bioventures), followed by customized production of DNA, proteins, etc. (78 bioventures), bioinformatics (41 ventures), and reagents and consumables development (38 bioventures).

Australia's biotechnology and bio-industry

In its 2003 global biotechnology census, the consultancy firm Ernst & Young ranked Australia's A$12 billion biotechnology and bio-industry as number one in the Asia-Pacific region and sixth worldwide. Australia accounts for 67 per cent of public biotechnology revenues for the Asia-Pacific region.

The Australian government gave a boost to the bio-industry by providing nearly A$1 billion in public biotechnology expenditure in 2002–2003. There were around 370 companies in Australia in 2002 whose core business was biotechnology – an increase from 190 in 2001. Human therapeutics made up 43 per cent, agricultural biotechnology 16 per cent and diagnostics companies 15 per cent. Over 40 biotechnology companies were listed on the Australian stock exchange (ASX) and a study released by the Australian Graduate School of Management reported that an investment of A$1,000 in each of the 24 biotech companies listed on the ASX between 1998 and 2002 would have been worth more than A$61,000 in 2003 – an impressive 150 per cent return. During the same period, shares in listed Australian biotechs significantly outperformed those of US biotechs, and the overall performance of listed Australian biotech companies was higher than that of the Australian stock market as a whole.

Over A$500 million was raised by listed Australian life science companies in 2003, and the ASX health-care and biotechnology sector had a market capitalization of A$23.4 billion in 2003, up 18 per cent on 2002. There has been a maturing of the Australian biotechnology sector, with greater attention paid to sustainable business models and the identification of unique opportunities that appeal to investors and partners. The industry is supported by skilled personnel – Australia is considered to have a greater availability of scientists and engineers than the United Kingdom, Singapore or Germany.

Australia is ranked in the top five countries (with a population of 20 million or more) for the number of R&D personnel. In terms of public expenditure on R&D as a percentage of GDP, it outranks major OECD countries, including the United States, Japan, Germany and the United Kingdom (Australian Bureau of Statistics, 2003). For biomedical R&D, Australia is ranked the second most effective country – ahead of the United States, the United Kingdom and Germany – particularly with respect to labour, salaries, utilities and income tax. Australia is ranked third after the Netherlands and Canada for the cost competitiveness of conducting clinical trials.

Australian researchers indeed have a strong record of discovery and development in therapeutics. Recent world firsts include the discovery that *Helicobacter pylori* causes gastric ulcers, and the purification and cloning of three of the major regulators of blood cell transformation – granulocyte colony-stimulating factor (GCSF), granulocyte macrophage colony-stimulating factor (GMCSF) and leukaemia inhibiting factor (LIF). Australia is cementing its place at the forefront of stem cell research with a transparent regulatory system and the establishment of the visionary National Stem Cell Centre (NSCC). An initiative of the Australian government, this centre draws together expertise and infrastructure; in 2003 it entered into a licensing agreement with the US company LifeCell.

Strong opportunities exist in areas such as immunology, reproductive medicine, neurosciences, infectious diseases and cancer. There are also opportunities for bioprospecting given that Australia is home to almost 10 per cent of global plant diversity, with around 80 per cent of plants and microbes in Australia found nowhere else in the world. Although 25 per cent of modern medicines come from natural products, it is estimated that only 1 per cent of plants in Australia have been screened for natural compounds.

Australia is the most resilient economy in the world, has the lowest risk of political instability in the world and possesses the most multicultural and multilingual workforce in the Asia-Pacific region. Its geographical location has not been a deterrent to the establishment of partnerships. According to Ernst & Young's 2003 "Beyond Borders" global biotechnology report, Australia had 21 cross-border alliances in 2002 – more than France and Switzerland, and 18 more than its nearest Asia-Pacific competitor. All the major pharmaceutical companies have a presence in Australia and pharmaceuticals are the third-highest manufactures export for Australia, generating over US$1.5 billion. The largest drug-exploration partnership in Australian history, between Merck & Co., Inc. and Melbourne-based Amrad to develop drugs against asthma, other respiratory diseases and cancer, was valued at up to US$112 million (plus royalties) in 2003. It is therefore no wonder that the pharmaceutical industry in Australia, which has annual revenues of US$9.2 billion, is increasingly viewed by the main global players as a valuable source of innovative R&D and technology.

2

Medical and pharmaceutical biotechnology: Current achievements and innovation prospects

Medical biotechnology – "red" biotechnology – may have its troubles, but at least most people worldwide favour developing new treatments, methods of diagnosis and prevention tools (e.g. vaccines). In the late 1970s, when the golden era of medical biotechnology started, the genes for proteins or polypeptides that worked or could work as drugs were cloned in microbial and/or animal cells, and the proteins were produced in bioreactors. Human insulin, human and bovine growth hormones, epidermic growth factor, erythropoietin, interferons, anti-haemophilic factors, anti-thrombotic agents (recombinant streptokinase and tissue-plasminogen activator), anti-hepatitis A and B vaccines, etc., have been produced in this way and successfully commercialized, as have mono-clonal antibodies that fuelled and transformed the diagnosis of pathogens and diseases.

Genomics, drug discovery and drug improvement

It is often stressed that many currently used medicines have only relative efficiency. For instance, anti-depressants are not effective among 20–50 per cent of patients, beta-blockers fail in 15–35 per cent of those treated, and one in five or even three people suffering from migraine cannot find an effective medicine to alleviate the pain (Mamou, 2004e). It is therefore expected that personalized medical care with drugs that take account of an individual's genetic make-up will improve the situation. Thus, at the

17

beginning of the annual report published by Burrill & Company – a Californian bank specializing in the funding of biotechnologies – Steven Burrill, its chief executive officer, predicted that "the era of a personalized medicine will generate a market characterized by a small volume per each drug, but the range of products developed for each therapeutic target will be much wider than presently". There is therefore a firm belief in the effectiveness of a future individualized medicine, which could be regenerative – tissue or even organ replacement – or preventive – for example, it would be possible to anticipate the occurrence of a cancer, rather than to have to try to cure it (Mamou, 2004e).

Although acknowledging some breakthroughs (the drug Gleevec has proved its efficacy against chronic myeloid leukaemia, and Genentech, Inc.'s Avastin can starve tumours by blocking the development of new blood vessels), analysts emphasize that the transition toward a new therapeutic era is quite slow. In the United States, the US$250 billion invested in biotechnologies from the late 1960s up to 2003 had a rather low return: out of the 200 best-selling drugs worldwide, only 15 per cent are derived from research and development (R&D) in the life sciences. According to data provided by the US Food and Drug Administration (FDA), in 1996, out of 53 drugs approved for sale worldwide, 9 were derived from biotechnology; in 2000, the figures were 27 and 6, respectively; and in 2003, 21 and 14. Most biotechnology companies continue to spend money on research that does not lead to marketable products. For instance, after 16 years of research into gene therapy and spending US$100 million, Vical has not found a marketable drug (Mamou, 2004e).

Therefore, deciphering the sequence of a gene and of the whole genome of an organism sounds like an attractive short cut, and genomics caught the attention of both the public and the stock markets during the last years of the twentieth century. Many new genes have been discovered, each implying the existence of at least one new protein that might have some therapeutic value.

For instance, an international scientific consortium comprising 58 institutions and laboratories announced the sequencing of almost the whole genome of the rat (*Rattus norvegicus*) in the *Nature* issue dated 1 April 2004. Rats, which come from Central Asia, have been widely used as laboratory animals in biological, medical and pharmaceutical research for the past century and half. This animal species became the third mammal after the human species and the mouse whose genome has been deciphered. The rat's genome is made up of 2.75 billion nucleotide pairs, which is intermediary in size between the human genome (2.9 billion nucleotide pairs) and that of the mouse (2.6 billion nucleotide pairs) (Nau, 2004c).

The consortium's work shows that 90 per cent of the genes in the rat's genome have their equivalents in the human and murine genomes; this similarity is interpreted as the three species having a common ancestor 20 million years ago. Other genes found in the rat's genome are absent in the other two mammalian species: these genes are involved in the production of pheromones, immune system processes and proteolysis; they are also related to detoxification mechanisms. This is an interesting finding because rats are frequently used to study the potential toxicity for humans of pharmaceuticals and chemicals. For economic reasons, the international consortium does not intend to pursue the in-depth study of the rat's genome (Nau, 2004c). Genome sequencing in mammals is being carried out on such species as chimpanzee, macacus, dog, bovine cattle and opossum.

On 20 April 2004, a team of 152 researchers working in 67 laboratories and scientific institutions in 11 countries (Australia, Brazil, China, France, Germany, South Africa, South Korea, Sweden, Switzerland, the United Kingdom and the United States), and coordinated by Takashi Gojobori (National Genetics Institute of Japan) and Sumio Sugano (University of Tokyo), announced that they had identified and described in a detailed manner 21,037 human genes. This group of researchers was created in 2002 under the name of "H-invitational" and their work is a follow-up to the sequencing of the human genome and its overall mapping in 2001. They published their results on the Internet in the free-access journal *PLoS Biology*, edited by the Public Library of Science. By so doing they wish to offer their results to the international scientific community (Nau, 2004d).

Starting from the human genome sequencing data, they identified the initial and final sequences of each of 21,037 genes out of the 30,000–40,000 that constitute the human genome. Their objective is to extract as much information as possible about the nature of these genes, their location and their functions, as well as their implications in a pathological process. This is an important step toward the elucidation of gene function, i.e. functional genomics, according to the French team who participated in this work – the National Centre for Scientific Research's Genexpress (Nau, 2004d).

However, genomics needs to be backed up with proteomics, transcriptomics, glycomics (to identify the carbohydrate molecules, which often affect the way a protein works) and metabolomics (studying the metabolites that are processed by proteins). There is even bibliomics and bioinformatics, which store and compare the sequences of genes and proteins, and search the published scientific literature to find connections between all of the above. But as Sydney Brenner, the 2002 Nobel Laureate for Physiology and Medicine, once observed, in biotechnology the one -omics that really counts is economics (*The Economist*, 2003a).

Most of the innovation in medical biotechnology, including the increasing reliance on genomics, has been done by small companies, so-called start-ups, in close collaboration with the universities. Across the United States, universities became hotbeds of innovation, as entrepreneurial professors took their inventions (and graduate students) off campus to set up companies of their own. Since 1980 (when the Bayh–Dole Act was enacted), American universities have witnessed a 10-fold increase in the patents they generate, spun off more than 2,200 firms to exploit research done in their laboratories, created 260,000 jobs in the process, and in 2002 contributed US$40 billion to the US economy.

I should also underline the strong support provided by public research institutions to US biotechnology. For instance, as part of their long-standing policy aimed at ensuring US pre-eminence in life sciences research and its applications, the National Institutes of Health distributed US$27.9 billion to researchers and universities in 2004. This budget was increased by the contributions of other ministries such as the Departments of Defense, the Interior and Agriculture. Such big public investment in basic research encourages private investors. Thus, in January 2004, despite the cautious approach of investors who bore the brunt of the drastic falls on the stock exchange in 2000, Jazz Pharmaceuticals – a one-year-old start-up – succeeded in raising US$250 million from private investors. In addition, funding associated with research to combat bioterrorism has helped many biotechnology companies specializing in immunology to survive (Mamou, 2004e).

If new drugs are to be discovered, exploiting genomics is one of the most likely routes to success. Some companies have understood this from the beginning. For example, Incyte, founded in 1991, and Human Genome Sciences (HGS), set up in 1992, both use transcriptomics to see which genes are more or less active than normal in particular diseases. But HGS saw itself as a drug company, whereas Incyte was until recently a company that sold its discoveries to others. As a result, in 2003, HGS had 10 candidate drugs in the pipeline, whereas Incyte had none (*The Economist*, 2003a).

The Icelandic company DeCODE Genetics is trying to use medical data on individuals to search for the genetic roots of disease. It has attracted controversy since July 2000, when bioethicists accused the firm of invading people's privacy and of not trying very hard to obtain people's consent before using their medical data. Three years later, the firm's methods were still viewed as shady, but DeCODE Genetics has found 15 genes implicated in 12 diseases, including the "stroke gene". The harmful form of this gene, which may cause plaque build-up in the arteries, is as much of a risk factor as smoking, hypertension and high cholesterol levels. Drugs to counter the gene are years away, and there is currently no way of knowing which form one has. But DeCODE Genetics' chief executive, Kari Stefansson, announced that a screening test could be ready in 2005.

The objective of, for instance, Perlegen and Sequenon is to connect genes to diseases and create drug-discovery platforms, e.g. through projects based on single-nucleotide polymorphisms (SNPs) or haplotypes. They study people's genomes only at the sites (such as SNPs) where variation is known to occur. Perlegen is using US$100 million of its start-up capital to record the genomes of 50 individuals (*The Economist*, 2003a). Proteomics has been picked up by Myriad, which formed a collaborative venture with the Japanese electronics firm Hitachi, and with Oracle, a US database company, to identify all the human proteins and to study their interactions by expressing their genes and examining their behaviour in yeast cells.

Genaissance, another haplotype company, is trying to connect genes not with diseases but with existing drugs, by examining how people with different haplotypes respond to distinct treatments for the same symptoms, e.g. the individual response to statins, which regulate the concentration of cholesterol in the blood – a US$13 billion market in the United States alone (Pfizer's statin, Lipitor, is the best-selling drug worldwide, and in August 2003 AstraZeneca was authorized by the US Food and Drug Administration to market its statin, Crestor, a formidable competitor to Lipitor). This kind of work may lead to "personalized medicine", i.e. to identifying an individual's disease risk and knowing in advance which drugs to prescribe. It would also help drug companies to focus their clinical trials on those people whose haplotypes suggest they might actually benefit from a particular drug. This approach will reduce the very high cost of testing drugs and will probably increase the number of drugs approved, since they could be licensed only for those who could use them safely. Presently, only about 1 out of every 10 molecules subjected to clinical trials is licensed. This drop-out rate explains part of the high cost involved in marketing a drug – US$500–800 million (*The Economist*, 2003a).

Another research trend in medical biotechnology is to modify the activity of proteins by acting on their genes. For instance, Applied Molecular Evolution has been able to obtain an enzyme 250 times more effective than its natural progenitor at breaking down cocaine. Genencor is designing tumour-destroying proteins as well as proteins that will boost the immune system against viruses and cancers, just like vaccines do. Maxygen has produced more effective versions of interferons alpha and gamma, to be tested on people, and is developing proteins that will behave like vaccines against bowel cancer and dengue fever (*The Economist*, 2003a).

X-ray crystallography of proteins is an efficient tool for unravelling their structure and can contribute to the design of a new drug. Thus Viracept, devised by Agouron (part of Pfizer), and Agenerase, developed by Vertex Pharmaceuticals (a biotechnology company), inhibit the HIV protease. Relenza, developed by Biota Holdings Limited, inhibits the neuraminidase of the influenza virus. Although a protein's three-dimensional structure can be deduced from its primary structure, i.e. the sequence of its amino-acids, it requires vast computing power. IBM's Blue Gene project is intended to solve the protein-folding problem, because the anticipated petaflop machine will be able to make 1 quadrillion calculations a second. To have a machine running at a quarter of a petaflop was considered an outstanding performance in 2004 (*The Economist*, 2003a).

Current achievements and prospects

Hepatitis C

The hepatitis C virus (HCV), which is spread mainly by contaminated blood, was not isolated and identified until 1989. In 1999, the most recent year for which global figures are available, HCV was believed to have infected some 170 million people worldwide; another 3 million are added every year. In most cases, the virus causes a chronic infection of the liver, which, over the course of several decades, can lead to severe forms of liver damage such as cirrhosis and fibrosis, as well as cancer. According to the World Health Organization (WHO), hepatitis C kills around 500,000 people a year. It is less deadly than HIV/AIDS, which claims more than 3 million lives annually out of some 42 million infected people. However, HCV's higher prevalence and longer incubation period, and the absence of effective drugs, mean that it is potentially a more lethal epidemic (*The Economist*, 2003d).

Effective new treatments for hepatitis C are not easy to develop, owing to the fact that the HCV is hard to grow in the laboratory and, until recently, the only animal "model" of the human disease was the chimpanzee, a species that it is impractical (and, many would argue, unethical) to use for industrial-scale research. However, new cell culture systems and mouse models have opened the way to further drug development. The NS3 protease of the HCV is a target, and scientists at the Schering-Plough Research Institute in New Jersey have begun clinical trials with an inhibitor of this viral protease. Vertex Pharmaceuticals has another anti-NS3 drug, VX-950, which blocks its target, at least in mice; it may be tested on humans (*The Economist*, 2003d).

Other substances aim at inhibiting the binding of HCV to liver cells in the first stage of infection. Among these is a compound from XTL Pharmaceuticals, which has been tested on 25 chronic sufferers. The drug is a monoclonal antibody designed to block the HCV's outer protein, E2, which the virus needs to attach to its target cells. In roughly three-quarters of the patients who received the compound viral levels dropped significantly, with no serious side-effects. As a result, XTL Pharmaceuticals was testing the drug on HCV-related liver transplant patients, hoping to prevent infection of the transplanted organ by hidden reservoirs of the virus. The company expected the results of the trials before the end of 2004 (*The Economist*, 2003d).

It is probable that a combination of drugs attacking the viral infection from different angles will be the most potent weapon. And, as with AIDS, success in drug-making will raise the thorny issue of access to effective drugs. Existing treatments, combining alpha-interferon and ribavirin (an inhibitor of viral replication) already cost US$20,000, which puts them beyond the reach of most of the world's infected people in developing countries. Future treatments, including a possible anti-HCV vaccine, may be more expensive and money will have to be found to pay for them when they arrive on the market (*The Economist*, 2003d).

Ebola fever

The Ebola virus is named after a tributary of the Congo river, close to the city of Yambuku (Zaire), where it was discovered in 1976 during an epidemic that affected 318 people and killed 280. It is one of the longest viruses known to date, consisting of a nucleic acid thread embedded in a lipid capsid. The incubation period of the disease varies from a few days to three weeks and the symptoms include fever, intense abdominal pain and haemorrhagic diarrhoea with liver and kidney dysfunction. The virus is transmitted through direct contact with contaminated blood, saliva, vomit, faeces or sperm; infected people should be put in quarantine. The haemorrhagic fever caused by the virus infection results in the death of 80 per cent of patients within a few days. Over the past few years, several of these fulminant epidemics have occurred simultaneously in the Democratic Republic of Congo (DRC) and Gabon, thus making infection by the Ebola virus a major public health priority for these countries. It should be noted that the Ebola virus that causes havoc in Gabon and the DRC belongs to the most virulent of the four subgroups known, the Zaire subgroup (Nau, 2003b).

Researchers from the French Research Institute for Development (IRD), together with researchers from the International Centre of Medical Research in Gabon, the World Health Organization, the US Wildlife Conservation Society, the Programme for the Conservation and Rational Use of Forest Ecosystems in Central Africa (a non-governmental organization in Gabon), the South African National Institute for Communicable Diseases Control and the US Center for Diseases Control, have been studying the virus since 2001 in the west of Central Africa. They assume that human epidemics caused by the virus originate from two successive waves of contamination: a first wave moves from the virus reservoir to some sensitive species, such as mountain gorillas, chimpanzees and wild bovidae; then a second wave infects humans through the carcasses of animals killed by the virus. According to the epidemiological data collected during the human epidemics that occurred between 1976 and 2001, each epidemic evolved from a single animal source and then spread through contact between individuals. However, a study carried out between 2001 and 2003 in Gabon and the DRC suggests the existence of several distinct and concurrent epidemic chains, each one originating from a distinct animal source. In addition, genetic analyses performed on patients' blood samples have shown that these chains stemmed not from a common viral strain but from several strains (Leroy et al., 2004).

On the other hand, the counting of carcasses found in the forests and the calculation of the indices of the animals' presence (faeces, nests and prints) have revealed an important increase in mortality among some animal species before and during human epidemics. Gorilla and wild bovidae populations halved between 2002 and 2003 in the Lossi reserve in the Congo, and the population of chimpanzees decreased by 88 per cent. Hundreds or even thousands of animals would have died during the recent epidemics in the region. It was verified that the decline in animal populations was due to infection by the Ebola virus. Genetic analysis of samples taken from the carcasses has shown the presence of several strains of the virus, as in humans (Leroy et al., 2004).

In conclusion, epidemics caused by the Ebola virus among apes result not from the propagation of a single epidemic from individual to individual, but rather from massive and simultaneous contamination of these primates by the reservoir animal in particular environmental conditions. Human contamination occurs in a second stage, generally through contact with animal carcasses. Consequently, the finding of infected carcasses can be interpreted as presaging a human epidemic. Such detection of animal carcasses, followed by a diagnosis of infection by the Ebola virus, would allow the development of a programme to prevent and control transmission of the virus to humans before an epidemic occurs. This would increase the probability of mitigating these epidemics or even avoiding them all together (Leroy et al., 2004).

A vaccine against the Ebola virus, consisting of an adenovirus into which the genes encoding the proteins of the Ebola virus have been transferred, has been made by the biotechnology company Vical. These viral proteins will induce the synthesis of antibodies against the Ebola virus in infected people. Because the vaccine does not contain any virus-derived structure, it is theoretically harmless. On 18 November 2003, the US health authorities announced a first clinical trial aimed at studying the innocuousness and efficacy of the vaccine. The US National Institutes of Health (NIH) indicated that the first phase of the clinical trial would involve 27 volunteers aged 18–44 years, of whom 6 would receive a placebo and 21 the vaccine in the form of three injections over a two-month period. The volunteers would be under medical supervision for a whole year. The clinical trial was a follow-up to experiments carried out on monkeys for three years by Gary Nabel at the Vaccine Research Center of the National Institute of Allergy and Infectious Diseases. These experiments led to the complete immunization of the animals (Nau, 2003b).

According to NIH director Anthony Fauci, an effective vaccine against the Ebola fever/virus would not only protect the most exposed people in the countries where the disease is naturally prevalent, but also deter those who might use the virus in bio-terrorist attacks. In addition to the vaccinia virus and anthrax, US specialists who fight bio-terrorism have been concerned for years about the possible use of pathogens that cause haemorrhagic fevers, and particularly the Ebola virus (Nau, 2003b).

Ribonucleic acid (RNA) viruses

Human immunodeficiency virus

The human immunodeficiency virus (HIV) that causes AIDS (acquired immunodeficiency syndrome) shows great genetic variability and is particularly virulent, probably because of its recent introduction into human populations. It has the potential to evolve very rapidly at the level of a population or an individual, owing to its mutation rate being one of the highest in living beings and to its capacity to recombine. This is a major obstacle to the production of an effective vaccine. Choisy et al. (2004) of the joint research unit of the French Research Institute for Development, the National Centre for Scientific Research and the University of Montpellier II, which is devoted to the study of infectious diseases following evolutionary and ecological approaches, in collaboration with the University of California, San Diego, and the University of Manchester,

United Kingdom, have tested the adaptive mechanisms of several HIV strains at the molecular level. They have studied and compared the evolution of three major genes of the HIV genome – *gag*, *pol* and *env* – in several subtypes of HIV.

The genes *gag*, which codes for the capsid proteins, and *pol*, which encodes the synthesis of the virus replication key enzymes, are very stable and conserved in all subtypes. By contrast, the gene *env*, which codes for the proteins of the external envelope of the virus (the targets of the immune system), would contain sites that are selected positively. Mutations of this gene have a selective advantage because they would result in the diversification of the expressed proteins – which would not be recognized by the antibodies. However, these proteins must conserve their vital function of adhesion of the virus particle to the membrane of host cells (CD4 cells of the immune system). This would mean that two opposed selection forces would operate on *env*, one toward conservation and the other toward diversification (Choisy et al., 2004).

The French researchers have confirmed a theoretical model proposed by US scientists in 2003, that the HIV uses very large complex sugar molecules to escape from the host's immune system. These sugars would create a kind of "shield" on the virus surface that prevents the fixation of human antibodies, without hindering the role of the envelope proteins in sticking the virus to its host cell. This finding applies to all tested HIV subtypes. It could lead to the development of new drugs against HIV/ AIDS and eventually to a candidate vaccine against all HIV strains. More research will be carried out to check the validity of these preliminary results and to make in-depth studies of the variability of SIV strains among primates, from which HIV strains have evolved (Choisy et al., 2004).

SARS virus

The SARS (Severe Acute Respiratory Syndrome) epidemic probably originated in the Guangdong province in early 2002 and spread to 28 countries. This disease is caused by a corona virus, whose genome is made up of 29,736 nucleotide pairs (the sequence was published on 13 April 2003 – an impressive achievement). Edison Liu and colleagues at the Genomics Institute in Singapore compared the genome sequences of corona viruses isolated from five patients with those of viruses studied in Canada, the United States and China. The researchers concluded that the virus was relatively stable compared with other RNA viruses (*The Lancet*, 9 May 2003). There were differences between the genome sequences of the viruses isolated from patients in Hong Kong and those of the viruses isolated from patients in Beijing and Guangdong. Such variation is useful in the study of virus dissemination and epidemiological follow-up.

Chinese scientists published their analysis of the evolution of the SARS virus in January 2004 in the journal *Science*. A consortium of researchers in Guangdong Province, Shanghai and Hong Kong, led by Guoping Zhao of the Chinese National Human Genome Centre, has shown that, as the virus perfected its attack mechanism in humans, its potency soared. Early on, it was able to infect only 3 per cent of people who came into contact with a patient; a few months later, the infectivity rate was 70 per cent (Wade, 2004a). Based on virus samples taken from Chinese patients in the early, middle and late stages of the epidemic, the analysis revealed the increasing infectivity of the virus as a result of evolution at the molecular level; it was therefore better to control the virus at a very early stage when the infection rate is lower. The Chinese researchers studied the SARS virus spike protein, which enables the virus to enter a cell. They found that the gene controlling the design of the spike protein mutated very rapidly in the early stages of the epidemic, thus producing many new versions of the spike protein. The new versions were maintained, an instance of positive selection pressure (Wade, 2004a). In the later stages of the epidemic, the sequence of the gene did not change, as if the spike protein had reached the perfect design for attacking human cells. The gene was under negative selection pressure, meaning that any virus with a different version was discarded from the competition. The evolution of the spike protein from the animal host of the virus to acquiring the ability to attack human cells began in mid-November 2002 and was complete by the end of February 2003, a mere 15 weeks later. Another gene that played a key role in the replication of the virus remained stable throughout the period when the virus was successfully switching from its animal to its human host (Wade, 2004a).

The study was praised by Kathryn Holmes, an expert on SARS-type viruses (corona viruses) at the University of Colorado, for its speed, the foresight in saving specimens from the critical early stages of the outbreak and its epidemiological analysis at the molecular level. Holmes stressed that this kind of evolution will occur in the future, referring to other pathogens that have moved from animal to human hosts. The SARS virus had probably infected humans many times before, but had failed to establish itself until 2002, when one of its constantly mutating versions succeeded in infecting humans (Wade, 2004a).

In June 2003, as a result of X-ray diffraction studies, Rolf Hilgenfeld and his colleagues at the University of Lübeck, Germany, published the structure of a proteinase that plays a key role in the replication of the SARS virus (Hilgenfeld et al., 2003). This proteinase is present in two strains of the corona virus, one that causes SARS in humans and the other that infects pigs. The biotechnology company Eidogen had also published the structure of this proteinase. Starting from the structural model developed by Hilgenfeld and his colleagues, German researchers suggested that a proteinase inhibitor (AG7088) that Pfizer had tested against the virus causing colds could be a good starting point for designing inhibitors of the SARS virus proteinase.

According to data provided by WHO, the SARS virus infected more than 8,300 people, of whom more than 700 died. Many researchers are of the opinion that the virus had been latent in an animal species before infecting humans. Yuen Kwok-Yung, a microbiologist at the University of Hong Kong Centre for the Control and Prevention of Diseases, and his team focused their research on exotic animals sold as food delicacies in a Guangdong market. They bought 25 animals from 8 different species; they isolated the corona virus in 6 civets and found the virus in another 2 species. The virus isolated from the civets was almost identical to that isolated from patients suffering from SARS, the difference being that its sequence had 29 extra nucleotides. Later, Henry Niman of Harvard University discovered that, out of 20 patients, only 1, originating from Guangdong, was infected by a SARS virus that conserved the 29 extra nucleotides. This patient might have been one of the first humans to be infected before the virus shed the extra nucleotides. Peter Rottier of Utrecht University hypothesized that the loss of these extra nucleotides rendered the virus infectious to humans.

Other researchers have been more cautious. For example, the WHO's virologist Klaus Stöhr stated that the animals that harbour the virus are not necessarily its reservoirs, and that a large number of animals should be screened before concluding that the civet is the virus reservoir. Moreover, it is not clear why some people harbour the virus without becoming ill and why children are not greatly affected by this disease. There are several methods to detect antibodies and viral material, but it has not been possible to identify the virus during the early days of infection, when patients feel well but are starting to spread the SARS virus.

Scientists at the US National Institutes of Health are working on the production of vaccines based on killed or attenuated viruses, proteins or DNA. On the other hand, some scientists are concerned about the eventual worsening of the disease as a result of vaccination, owing to interaction with the immune system, as occurred with a vaccine against the corona virus that causes feline infectious peritonitis. The US National Institute of Allergy and Infectious Diseases (NIAID) has allocated US$420,000 for developing a vaccine against SARS, using an adenovirus as a vector.

Two biotechnology companies have announced their intention to develop therapies against SARS. One of them, Medarex, signed an agreement with the Massachusetts Biological Laboratories of the University of Massachusetts School of Medicine with a view to producing a therapeutic human monoclonal antibody. The other company is Genvec, which is collaborating with NIAID on the vaccine development project. On the other hand, two days after the publication of the genome sequence of the SARS virus, Combimatrix (part of Acacia Research) developed a micro-array chip of the virus, which is dispatched free of charge to key research centres. In Germany, Artus, which is collaborating with the Bernhard Nocht Institute of Tropical Medicine, has used the data on the virus genome sequence to develop the first diagnostic test, with a view to selecting targets for drugs against the virus.

Avian flu virus

Bird or avian flu virus first demonstrated an unprecedented ability to infect humans in 1997 in Hong Kong. On the advice of influenza specialists, the government ordered all 1.4 million of the territory's chickens and ducks to be slaughtered. Although 18 people were infected, of whom 6 died, the swift culling eliminated a potentially much greater disaster (Elegant, 2004).

By early 2004, the avian flu virus strain H5N1 had infected mainly chickens – 80 million of them had died from the flu or been killed and had been burnt or bagged and buried alive, in an effort to keep the disease from spreading. Up to February 2004, the virus had killed 22 people. Most of them had probably come into contact with the birds' faeces or perhaps inhaled infected dust blown by flapping wings. Health officials were concerned less about the danger to farm workers than to the wider public. Should strain H5N1 acquire the ability to pass from human to human, instead of only from bird to human, the consequences would be much more dramatic than the SARS epidemic (Guterl, 2004).

The geographical spread of the 2004 avian flu outbreak was, of course, a crucial reason why eradicating it was so much harder than in 1997. The flare-up of avian flu was confirmed in 10 Asian countries and territories: China, Japan, South Korea, Taiwan, Thailand, Vietnam, Indonesia, Cambodia, Laos and Pakistan. Many scientists believe that migratory wild birds, which can carry many viruses without showing symptoms of disease, were most likely the agents of the initial outbreak of the disease. Other factors, such as the transport of infected chickens across borders (legally and illegally), as well as the reluctance of governments to acknowledge the existence of the outbreaks, came into play and caused the greatest alarm (Elegant, 2004).

Many analysts and media professionals have stressed the official stonewalling and reluctance to acknowledge mistakes or to act swiftly to take preventive measures. For instance, although the head of the Bogor Institute of Agriculture's Faculty of Animal Husbandry suspected an outbreak of avian flu as early as August 2003, Indonesian officials did

not admit until 25 January 2004 that the country was facing a major outbreak. In Thailand, the idea of a bird flu epidemic was dismissed as an
exaggeration that would damage the country's poultry exports (Thailand
is the world's fourth-largest exporter of poultry) and harm farmers and
workers involved in the chicken industry. Avian flu has now been de-
tected in nearly half of Thailand's 76 provinces, and almost 11 million
birds have been culled across the country. China, the world's second-
biggest poultry producer, has been the source of many of the major flu
viruses to hit the world in the past 100 years, owing to its vast population
of chickens and ducks living in close proximity to each other and their
human owners. After weeks of hesitation and ambiguous information,
and despite its previous experience with handling the SARS epidemic,
on 27 January 2004 the central government finally acknowledged that
the bird flu outbreak had reached China. The Ministry of Agriculture,
in a radical shift towards an open approach to tackling the problem,
demanded that all outbreaks be reported within 24 hours. By the end of
January 2004, officials had confirmed outbreaks in Hunan and Hubei
provinces in central China, in addition to the cases reported across eastern China: Anhui and Guangdong provinces were potential hotspots, as
well as Kangqiao, a suburb of Shanghai. The lesson to be drawn from
both the SARS and avian flu epidemics is that, when it comes to fighting
highly contagious diseases, nothing is more important than decisive government intervention and transparency (Elegant, 2004).

However, rigorous government intervention is not enough to contain
bird flu. In Asia's countryside, almost everyone raises chickens or ducks,
and animals and humans live so closely together that the spread of viruses seems almost unavoidable. Health and agriculture experts believe
that livestock husbandry practices are at the heart of the bird flu crises,
especially in South-East Asia. Changing these practices is a great challenge, but the economic consequences of the epidemic should also be a
warning to the countries involved to make the necessary changes.

In this respect, although the US and Chinese economies were likely to
expand strongly in 2004 – Merrill Lynch forecast that Asian economies,
excluding Japan's, would grow 6.1 per cent in 2004, and US investment
house T. Rowe Price expected corporate profits across Asia to surge
about 15 per cent – the Asian Development Bank's assistant chief economist warned that, if avian flu were not curtailed soon, it "could cost
the region tens of billions of dollars" (cited in Adiga, 2004). Thailand's
US$1.25 billion poultry industry was set to be devastated as exports
to many markets were temporarily stopped. And tourism may also be
threatened, although the SARS epidemic had much more drastic consequences. In 2003, SARS-related lost business revenues were estimated
to amount to US$59.0 billion in Asia (China, US$17.9 billion; Hong

Kong, US$12.0 billion; Singapore, US$8.0 billion; South Korea, US$6.1 billion; Taiwan, US$4.6 billion; Thailand, US$4.5 billion; Malaysia, US$3.0 billion; Indonesia, US$1.9 billion; the Philippines, US$600 million; Vietnam, US$400 million) (Adiga, 2004).

However, according to Daniel Lian, a Thailand analyst at Morgan Stanley, even if Thailand's poultry exports were to fall to zero for the first quarter of 2004, avian flu would reduce the country's total exports by only 0.4 per cent in 2004. Unless the flu spread, this analyst expected Thailand's projected 8 per cent growth for 2004 to drop by only 0.11 of a percentage point. Other bankers and investors considered that the economic prospects for Asia looked bright. The ultimate economic impact of the avian flu epidemic will depend not only on how quickly governments control the spread of the disease but also on how deftly they manage international perceptions of the threat (Adiga, 2004).

The world has been afflicted by human flu epidemics before; for example, the 1918 Spanish flu pandemic claimed up to 50 million lives. After Jonas Salk's efforts to improve influenza vaccines in the 1940s, 46,000 people died in the 1968 Hong Kong flu pandemic. In 1976, swine flu vaccine produced polio-like symptoms, and in 1997 avian flu claimed its first human victims, although it did not spread among people. The deadliness of the avian flu strain H5N1 caught the attention of scientists: whereas the 1918 flu pandemic killed up to 4 per cent of those infected, and in 2003 SARS killed 11 per cent, in 1997, of the 18 people infected with the H5N1 flu strain in Hong Kong, 6 died, i.e. the mortality rate was 33 per cent. What really concerned the scientists was not only the mortality rate, but the persistence of the avian flu strain in trying to cross the species barrier. They feared that, sooner or later, it would infect a human and, through random mutation, adopt a form that allowed human-to-human transmission. Or it could acquire this ability by swapping genes with another flu virus already adapted to humans. Or the same event could occur in a pig infected with both the avian flu virus and a human flu virus (Guterl, 2004).

There is no human vaccine for avian flu. Under the WHO's aegis, laboratories in the United States and United Kingdom have begun developing a vaccine seed from viral specimens sampled from the 2004 outbreak, but this development and testing could take six months. Nine pharmaceutical companies make more than 90 per cent of the influenza vaccine in the world. Diverting those resources to stockpile an avian flu vaccine would take time and would disrupt the supply of human flu vaccine. The deadly nature of the bird flu virus presents another drawback. Flu vaccines are generally prepared from viruses cultured in fertilized hen eggs, but the H5N1 virus is as lethal to the embryo inside an egg as it is to adult birds (Elegant, 2004).

Vaccine developers therefore have to use a different vaccine production process, "reverse genetics". In 1992, Peter Palese of Mount Sinai Hospital in New York developed a technique for replicating RNA viruses through the replication of RNA into DNA and back again. The technique was refined by several research teams over the following years. Virologists Robert Webster and Richard Webby from St. Jude Children's Research Hospital in Memphis, Tennessee, used this "reverse genetics" technique to produce an experimental vaccine for the 1997 bird flu strain H5N1. First, they clipped off genetic material in order to make the virus no longer contagious. They then combined RNA from the virus with that of another known virus, called the "master seed virus", turned the whole lot into DNA, and tested it to make sure that the combination worked. They replicated the DNA back to RNA and obtained a genetically engineered flu virus. This result did not raise much interest. Vaccine companies were concerned about licensing arrangements with MedImmune – the firm that holds the patents for "reverse genetics" – and regulators were wary of the genetic engineering approach, which could entail years for regulatory approval (Guterl, 2004).

However, the rapid spread of the virus strain H5N1 changed the situation. The WHO's Global Influenza Program director, Klaus Stöhr, demanded a teleconference on 4 February 2004 with regulators and vaccine makers, who expressed their willingness to work together and solve the numerous logistical problems relating to large-scale vaccine distribution. Webster and Webby, after obtaining samples of the saliva of two Vietnamese patients infected with the H5N1 strain, were trying at St. Jude Children's Research Hospital to attach the genes' coding for two surface proteins of the H5N1 virus envelope to the master seed virus, and thus develop a virus with the ability to trigger the human immune system. They expected to produce a vaccine within a few months (Guterl, 2004).

Scientists are therefore optimistic that a vaccine could be delivered before the bird flu outbreak becomes a pandemic. If the pandemic spreads before a vaccine is ready, physicians will have to rely on the drugs available for standard flu. Unfortunately, strain H5N1 is already showing resistance to amantadine, a cheap and widely available drug. A more expensive drug, Tamiflu, will remain effective against the viral strain, but supplies are limited and it may be out of the reach of most Asian patients. It also seems that making a vaccine against flu, even a new strain of avian flu, is easier than creating one for an entirely new disease, as is required for HIV/AIDS and SARS. Although H5N1 has proved mutable, the WHO is confident that a vaccine will remain effective even if the virus undergoes a genetic shift that enables it to spread easily among humans. However, the immediate approach is to stop the dissemination of the virus among birds before a human vaccine becomes available; because, as stated by Yi Guan (a SARS and avian flu expert at the University of Hong Kong), "once this virus can spread from human to human, region to region, it is too late" (Elegant, 2004).

By March 2004, avian flu had disappeared from the headlines, but it was not because the threat had receded. The risk of the H5N1 avian flu strain mutating into a human-transmissible strain remains as high as ever. Different strains of bird flu or influenza have reached three countries with reliable animal health reporting systems – Japan, Canada and the United States – but the authorities were unable to stop it from spreading. On 3 April 2004, Canada reported that avian flu had spread outside its quarantined "hot zone" and infected at least 2 workers, with another 10 suspected cases (Bonte-Friedheim and Ekdahl, 2004).

Before the H5N1 outbreaks were reported in Cambodia, China, Indonesia, Japan, Laos, Thailand and Vietnam, avian flu had never infected so many birds over such a large area. Earlier outbreaks of a less pathogenic avian flu – in Pennsylvania in 1883–1885 and Italy in 1999–2000 – were localized, but even so they required stringent controls over several years to extinguish. According to Klaus Stöhr, the WHO's Global Influenza Program director, "given how far H5N1 avian influenza has spread, the world will be on the verge of a pandemic for at least a year, more likely two years" (Bonte-Friedheim and Ekdahl, 2004).

It is true that living in close proximity to animals exposes humans to animal viruses against which we have neither antibodies nor genetic immunity. Often, only one small mutation can change a virus's animal receptor into a human receptor. As a result, some of the biggest scourges have been diseases of animal origin. In the case of the H5N1 strain, the pandemic risk arises if a person is simultaneously infected by both the avian flu virus and a human flu virus. The two viruses could exchange genes, producing a recombinant avian–human virus that is effectively transmissible from human to human. However, according to the available data all people infected by H5N1 contracted the virus from birds. There is no confirmed evidence of human-to-human transmission yet (Bonte-Friedheim and Ekdahl, 2004).

There are several phases in the emergence of a typical pandemic: first, the virus becomes endemic in its animal host; then it crosses the species barrier to humans; and finally it becomes efficient at spreading from person to person. This possibility and the danger grow with the rise in international trade and travel. In the case of the SARS virus, analyses indicate that it moved from its original animal host to humans by mid-November 2002. Only in early 2003 did it develop an efficient mechanism of spreading between people. Once this was achieved, the virus spread around the world within weeks; it spread to Toronto via Hong Kong and killed Canadians before the medical authorities even knew there was a major problem. Airborne influenza viruses are much more easily spread than SARS, partly because still-healthy carriers shed large amounts of virus. That is why the basic containment measures – isolation and quarantine – that were very effective in the case of SARS would not be as efficient if we were dealing with a recombinant avian–human flu virus (Bonte-Friedheim and Ekdahl, 2004).

Bonte-Friedheim and Ekdahl (2004) believe that governments should implement the following preparedness measures in order to protect their populations from a pandemic caused by a recombinant avian–human virus:

- contingency planning for health-care systems: a new influenza strain would rapidly exhaust health-care systems' resources, so plans to activate back-up facilities and staff should be drawn up;
- strengthening of flu vaccine programmes in order to create background protection as well as more capacity to mass produce a new type of vaccine if needed;
- stockpiling of anti-influenza drugs; current anti-viral drugs may not cure all cases of a new H5N1 avian influenza strain, but they could limit the infection and can protect health-care workers;
- compliance of surveillance in countries with H5N1 avian flu with the WHO's guidelines and procedures.

Chagas' disease

Chagas' disease, or American trypanosomiasis, affects some 20 million people in the tropical regions of Central and South America. It is caused by the flagellated protozoan, *Trypanosoma cruzi*, which is transmitted to humans by Triatomidae insects. Researchers at the French IRD Unit on Pathogenies of Trypanosomatidae and their collaborators at the National Institute for Health and Medical Research (INSERM) have studied the life-cycle of the blood parasite, its virulence and its role in the infection, so as to identify possible ways of preventing and controlling the disease. In addition to deceiving the immune system of the host, *T. cruzi* is found during its life-cycle in the form both of flagellated cells, which circulate and multiply in the bloodstream, and of non-flagellated cells, which are intracellular and give rise to the flagellated circulating cells. Both kinds of cell are able to secrete a protein, Tc52, that has enzymatic and immunosuppressive activities. This protein inhibits the production of interleukin-2 (IL-2) – a cytokine necessary to the proliferation of T-lymphocytes, and as such has an immunosuppressive action.

Mice have been immunized with this protein and then infected with *T. cruzi*. The result has been a decrease in the mortality rate during the acute stage of the disease, demonstrating the protective effect of Tc52.

When mutants of *T. cruzi* lacking the gene encoding Tc52 were used to infect mice, the mice were able to produce interleukin-2 normally while showing attenuated symptoms of the disease. The French researchers have identified the minimal sequence of the protein that causes the immunosuppressive effect. It is now possible to design chemotherapeutic strategies, i.e. the inhibition of the enzymatic activity of Tc52 by antiparasitic drugs, or vaccination protocols against the protozoan. Research is being carried out in collaboration with INSERM and the Laboratory of Immunology and Therapeutic Chemistry of the National Centre for Scientific Research on the identification of Tc52 receptors existing on macrophages and dendritic cells and on the synthesis of specific inhibitors of these receptors, thus paving the way for the development of new medicines against Chagas' disease.

Type-1 diabetes

Type-1 diabetes is the result of disorders in the immune system: for reasons not yet well understood, a combination of genetic and environmental factors induces the destruction of insulin-producing islet cells (Langherans islets) in the pancreas. Inflammation plays a role in starting this process. Yousef El-Abed at the North Shore-Long Island Jewish Research Institute suggested that a protein known as macrophage migration inhibitory factor (MIF) might make a logical therapeutic target. MIF signals to a network of inflammatory pathways. By the end of March 2004, at a meeting of the American Chemical Society, El-Abed announced he had found a compound that might halt inflammation by cutting the communication between MIF and inflammatory pathways. This compound, named ISO-1, binds to a particular site on the surface of MIF and prevents it from carrying out its inflammatory properties, in the laboratory at least (*The Economist*, 2004b).

Experiments were then carried out on a group of laboratory mice treated with a chemical called streptozotocin or ST2, which caused them to experience the high blood sugar levels that characterize diabetes. The mice not treated with ISO-1 all developed chemical-induced diabetes, whereas those that were treated with ISO-1 did not. El-Abed and his colleagues also investigated the drug's effectiveness on another group of mice that had been genetically engineered to develop type-1 diabetes. Of these, 20 out of 22 were protected. None of the treated mice suffered from any side-effects (*The Economist*, 2004b).

El-Abed thinks ISO-1 could be easily adapted for a human oral medicine within a few years. Test kits would have to be developed to find the children who could benefit most. The US researcher envisages a population-wide screening, like the infant heel-prick test that is used to detect phenylketonuria in newborns. Individuals with high levels of diabetes-related antibodies would be monitored and, when the levels indicated an impending attack on pancreas islet cells, the drug would be administered. Costly though such screening would be, it could save the US$100 billion a year that is spent treating diabetes in the United States alone; type-1 diabetes affects around 10 per cent of diabetes sufferers, but it accounts for about 40 per cent of treatment costs (*The Economist*, 2004b).

Autoimmune diseases

After talks with the US Food and Drug Administration (FDA), the Irish drug-maker Elan plc and the US biotechnology enterprise Biogen Idec expected to file for approval of their multiple sclerosis (MS) drug Antegren (natalizumab) in mid-2004 – one year earlier than anticipated. If approved, the monoclonal antibody could come to market in 2005 (rather than its previous target of 2006). Biogen Idec and Elan are collaborating equally on the development of natalizumab for MS, Crohn's disease and rheumatoid arthritis. The humanized monoclonal antibody prevents immune cells from leaving the bloodstream and migrating into chronically inflamed tissue by targeting the selective adhesion molecule (SAM).

The growing and hotly contested MS drug market is estimated by some analysts to be worth up to US$4 billion. Currently, it is dominated by the German drug-maker Schering (Betaseron), Switzerland-based Serono SA (Rebif), Israeli Teva (Copaxone) and Biogen (Avonex). Shares in Schering, Serono and Teva fell upon the news.

Celltech Group plc and Biogen Idec have entered into collaboration for the research, development and commercialization of antibodies to the CD40 ligand (CD40L) protein for the treatment of autoimmune diseases. The CD40L protein is a key regulator of antibody-mediated immune responses. Blocking the interaction between CD40L on T-cells and CD40 on B-cells has been shown to reduce excessive antibody production and may help restore a normal immune response in patients with a variety of autoimmune-related conditions, including rheumatoid arthritis, multiple sclerosis and inflammatory bowel disease. Celltech will be responsible for the identification and engineering of new high-affinity antibodies to CD40L and will bear all development costs until the end of phase 1 of the human safety testing. Following completion of phase 1, Biogen Idec has an option to co-invest in the ongoing development of products. In this case, the companies will jointly develop and commercialize products and will share costs and profits. Alternatively, if Biogen Idec does not exercise its option, Celltech may elect to take the programme forward independently, and continue to develop and market products on an exclusive, worldwide basis. Biogen Idec would then receive royalties based on the sales achieved by Celltech.

The fight against cancer

In the fight against cancer, aggressive treatments such as radiation and chemotherapy were often considered 30 years ago to be the only options for survival apart from surgery. Unfortunately, they routinely did patients as much harm as good by killing off all dividing cells, cancerous or healthy. Once a cancer is identified and a combination of drugs to combat it has been chosen, there is no guarantee that the drugs will work. That is because no two patients are alike. Subtle differences in people's genetic make-up often determine how well a cancer drug will be tolerated and how quickly it will be broken down in the body. Some individuals produce enzymes that can neutralize the more toxic side-effects of anti-cancer drugs, whereas others either lack such enzymes or have genes that make them more sensitive to the drugs' adverse effects. Researchers at Massachusetts General Hospital (MGH), for instance, found that changes in the gene coding for an enzyme in DNA repair can mean the difference between breast-cancer patients who can tolerate chemotherapy and those with a two-fold greater chance of experiencing a toxic reaction (*Time*, 21 June 2004, p. 58).

In 1970, a new field opened up: how to coax a person's immune system into fighting cancer. Ronald Levy, now chief of the oncology division at Stanford University School of Medicine, began his training in cancer immunology when he signed up for a two-year stint at the National Cancer Institute (NCI). Those were exciting times – President Richard Nixon had just declared his "war on cancer" and funds poured into research projects. The idea was to make the connection between laboratory science and clinical medicine (*The Economist*, 2002b).

Two research teams in the United Kingdom made some breakthroughs. Each healthy B-lymphocyte expresses a unique "marker" protein on its surface, which it uses for binding antigens. When B-lymphocytes detect a foreign substance in the blood, they divide and churn out the same protein; this then binds to the antigen, tagging the invader for destruction. In B-cell lymphoma, one cell divides endlessly, but this time without a purpose. In 1975, George and Freda Stevenson, of the University of Southampton, discovered that, in each patient, all malignant B-cells displayed the same "marker" protein, which could be used as a target. At the same time, César Milstein and Georges Köhler, of the Medical Research Council's Laboratory of Molecular Biology in Cambridge, worked out how to make monoclonal antibodies (they later received a Nobel prize). They fused two types of mouse cells – a tumour cell and a lymphocyte;

the new hybrid cell had the properties of both, and not only was it immortal but it could also produce antibodies forever (*The Economist*, 2002b).

When Levy realized the implications, it became clear that he not only had a way to identify and target all cancerous B-cells in one patient, but also would have the ability to create a potent customized therapy that would exploit a B-cell's unique marker protein. In 1975, Levy was hired by Stanford University School of Medicine and was eager to test his ideas in his own laboratory. After asking a few colleagues from around the world, including the Stevenson team, he settled on developing a patient-specific antibody able to latch onto the target directly. Although Levy was able to repeat his findings, many other researchers were not. There were two difficulties: in some cases, the human immune system reacted strongly against the foreign mouse molecules and neutralized the antibodies; and many antibodies were not aimed at the right targets and had little effect. It would take more than a decade to sort out these problems (*The Economist*, 2002b).

The promise of antibodies (or "magic bullets" as they were known at the time) boosted funding for biotechnology start-ups round the world, and helped Levy to co-found a company, Idec Pharmaceuticals, in 1985. But commercial success did not come easily and it was more than a decade before a modified version of the antibody was developed by Idec Pharmaceuticals and commercialized, though it was not customized for each patient. In 1997, the FDA approved Rituxan – the first monoclonal antibody for use against cancer and also the first lymphoma medication to hit the market in 20 years. It has since become a US$1 billion drug (*The Economist*, 2002b).

Many in the field give Levy credit for laying the foundation for Rituxan's success, but he also never gave up trying to make a personalized drug. Since 1988, he has led about a dozen small studies, testing a patient-specific cancer vaccine. Rather than offering protection against the disease, its aim is therapeutic, for those already afflicted. At least half the 200 or so people who received the medicine showed an immune response, and many stayed in remission for longer than the two years that are typical following chemotherapy. In a few cases, tumour shrinkage was observed, lasting four to five years. Several patients have remained disease-free for years after receiving the treatment. But, even if the vaccine eventually proves successful in large randomized clinical trials, can it be manufactured cheaply enough to be cost-effective? Levy remains as determined as ever to find a solution. For one thing, vaccines are only half as much trouble to make as antibodies. Rather than manufacturing the antibody, scientists isolate the "marker" protein and mix it with an immunity booster, hoping that the formula will coax the patient's own immune system into action against the cancerous cells (*The Economist*, 2002b).

Other researchers have carried out their own studies in the field. The closest to success looks like being Larry Kwak, now a principal investigator at the NCI's experimental transplant and immunology branch. After completing a series of small, successful studies, Kwak has conducted late-stage trials, which are usually the last step before receiving approval from the FDA. So far, no cancer vaccine – customized or otherwise – has reached the market. But, for B-cell lymphoma patients, the wait may be coming to an end. Data from trials by Genitope, a biotechnology company, were expected to be released by the end of 2004 and, if the FDA deems the study a success, approval could follow a year or so later – a quarter of a century after Levy successfully tested a monoclonal antibody against B-cell lymphoma in a patient who was 88 years old at the end of 2002 (*The Economist*, 2002b).

The following anti-cancer drugs are currently being used in the United States:

- Gleevec, approved by the FDA against chronic myeloid leukaemia and gastro-intestinal stromal tumours – blocks the key signals that cancer cells use to drive growth;
- Erbitux, a monoclonal antibody approved by the FDA against advanced colon cancer – blocks the epidermal growth factor, a key protein that cancer cells need to continue dividing;
- Iressa (gefitinib), approved for treating lung cancer – blocks the epidermal growth factor; and
- Avastin, the first drug approved by the FDA – starves tumours by blocking the development of new blood vessels (angiogenesis).

The following anti-cancer drugs were still in trials in mid-2004:

- Tarceva, also aimed at blocking the epidermal growth factor, was being tested for the treatment of both advanced lung cancer and pancreatic cancer;
- SU 11248 blocks tumours by interfering with their ability to generate blood vessels and by inhibiting enzymes essential for growth;
- BAY 43-9006, under study in kidney cancer, interferes with a newly discovered group of growth signals;
- Tarceva + Avastin could block distinct but critical growth pathways of cancer cells.

Most of the newly approved drugs work in only 10–30 per cent of patients, but in those patients tumours routinely shrink to less than half their original size. The number of new drugs that have been approved is small, their cost is high (at least US$20,000 per cycle), and progress is slow. According to the American Cancer Society, the five-year survival rate for all cancers in the United States was 63 per cent in 2003, up from

51 per cent in 1975. But it seems that most of that improvement was attributable to the effectiveness of anti-smoking campaigns, not to better drugs. However, the better researchers know a cancer, the better their chances of defeating it. As we know more about the behaviour of tumour cells at the molecular level, we are becoming convinced that the one-drug-one-cancer approach is not sufficient. Just as a multi-drug approach is used to attack HIV at different stages of its life-cycle, so too are cancer specialists beginning to treat tumour cells with combinations of drugs that can weaken a growing cancer by chipping away at its life-support systems. In coming years, doctors will think not of breast and colon cancers, but rather of the signalling pathways the cancer cell is using – growth factors, angiogenesis factors, etc. (*Time*, 21 June 2004, p. 58).

Antibiotics

Nowadays, more than 5,000 antibiotic substances are known. The problems of continuous production and stability have been solved, making antibiotics cheap and their supply secure. The compounds are produced in huge bioreactors holding up to 200,000 litres. Generally, the cultured strains of micro-organisms (e.g. *Streptomyces* and streptococci) are genetically engineered to guarantee high yields at high degrees of purity, by maintaining culture stability as long as possible. World production exceeds 30,000 tons, with a total market value of US$24 billion by the late 1990s (European Commission, 2002).

The first class of antibiotics, based on enzymatic activity, was followed in 1994 by a second, the peptide antibiotics. Since then, several other classes have been discovered and marketed: methicillins, vancomycins, aminoglycosides, macrolides, cephalosporins, quinolones, lipopeptides, glycopeptides. They are all based on a mere 15 compounds, such as the beta-lactams to which penicillin and the cephalosporins belong. All currently used antibiotics were introduced between 1940 and 1962; then, after a gap of 38 years, a new class of oxazolidinones followed in 2000. These function by blocking protein synthesis in bacteria.

Between 1990 and 1998, the number of reports of bacterial resistance increased from 30,000 to 50,000. The development of resistance was largely due to the fact that from 1962 until recently only modifications of existing antibiotic classes had been launched. Bacterial resistance to one product could more easily adapt to the whole class. In the United States alone, 50–60 per cent of nosocomial infections involve antibiotic-resistant bacteria, adding at least US$4.5 billion to health-care expenses. This threat induced the search for completely different compounds that attack bacteria through new mechanisms: fluoroquinolones, quinoprisitin, dalfoprisitin, linezolid, ketolides and glycylcyclines. Interesting compounds have been found in animals – e.g. various groups of anti-microbial peptides, such as magnainin from frogs.

An alternative is the search for specific genes in the sequenced genomes of major pathogenic bacteria. The idea is to decipher the sequence of a gene coding for key metabolic processes in the pathogenic microorganism, and then to engineer an inhibitor molecule. Experts estimate that bacterial genomics has already produced about 500 to 1,000 new broad-spectrum antibacterial targets. In addition, bacteriophages are making a comeback and can be of significant help in a few specific applications (European Commission, 2002).

Another promising approach has been the combining of beta-lactam antibiotics with a so-called "guardian angel" compound that neutralizes the beta-lactamase of the antibiotic-resistant bacteria. Clavulanic acid is one such compound, but it is ineffective in protecting the cephalosporins, which are frequently used in hospitals. The beta-lactamase inhibitors already developed have not progressed much into clinical use because of their high production cost (European Commission, 2002).

One of the research projects dealing with antibiotic synthesis and which received funding under the Fifth Research Framework Programme of the European Commission (1999–2002) – Towards New Antibiotics – aims at using genetically engineered micro-organisms to provide efficient routes for producing templates for modification into antibiotics and beta-lactamase inhibitors. It also seeks clean synthesis routes to currently used antibiotics in order to avoid costly and environment-unfriendly production procedures. It also intends to make possible the large-scale production of broad-spectrum beta-lactamase inhibitors.

Diagnostics

Diagnostic tests are a key area of medical biotechnology. Their rocketing development is illustrated by the fact that DNA profiling has become a cornerstone of forensic practice worldwide, less than 20 years after its invention in 1984, when Alec Jeffreys, a geneticist at Leicester University, discovered parts of the human genome that vary greatly between individuals. His laboratory developed a method of extracting these "short tandem repeats" (STRs) from DNA samples and making them visible with radioactive probes on X-ray film. Nowadays, the process is highly automated and can extract enough DNA from a minute biological sample, such as a speck of dandruff or saliva, to match against a DNA database. Tests are rapidly becoming faster to carry out, cheaper and more sensitive (Cookson, 2003).

The United Kingdom's National DNA Database, set up in 1995, is the world's largest, holding DNA profiles of 2.1 million people. Police in the

United Kingdom are allowed to take a sample from anyone who is suspected of, arrested for, or charged with a crime. The profile remains in the database whether or not the person is eventually convicted in court. Civil liberty groups are unhappy about the arbitrary nature of this policy. Nevertheless, a four-year, £182 million programme to expand the database to include the "entire active criminal population" was to be completed by April 2004. Some people, including Alec Jeffreys, wanted the national DNA database to encompass the whole population. "If everyone were on it, the unfairness of the present system would disappear", the geneticist stated (Cookson, 2003, p. 9). According to James Watson, the co-discoverer of the structure of DNA in 1953, "if everyone's genetic fingerprint were taken, it would take away our liberty to commit crime" (Swann, 2004, p. W3).

But the Home Office insists: "There is no government agenda to get everyone's DNA profile on the data-base. Sampling the entire population raises significant practical and ethical difficulties; we would have to consider the benefits it would bring and its compatibility with basic human rights" (cited in Cookson, 2003). Meanwhile, the existing database was bringing increasing rewards for law enforcement: according to the government's Forensic Science Service, which runs the database, in a typical month DNA matches are found linking suspects to 15 murders, 31 rapes and 770 motor vehicle crimes. At the same time, old crimes are being solved as the database grows. For instance, when John Wood was arrested in 2001 for stealing £10 of groceries, his DNA sample was fed into the database and matched the profile of a rapist who had attacked two girls in 1998. He pleaded guilty to the assault and is now serving a 15-year prison sentence (Cookson, 2003).

In the United States, the federal administration is facing fiercer opposition from groups such as the American Civil Liberties Union to its plan for a big expansion in the Combined DNA Index System – a joint programme between the FBI's National DNA Index System and state and local databases. The states vary greatly in the size and ambition of their DNA profiling activities. Virginia's database has 207,000 profiles and in the first nine months of 2003 there were 440 "cold hits", where DNA analysis of a crime-scene sample with no suspect matched a profile in the database. Automation seems to have eliminated human error from the profiling process and lawyers have learnt how to present statistics to juries without being accused of exaggeration (Cookson, 2003).

Since 1999 the UK national database has used a profile that combines 10 STRs. The chance of someone else sharing the same profile as the suspect is less than 1 in 1 billion, according to the Forensic Science Service. The United Kingdom is locked into this methodology (known as SGM Plus) for the foreseeable future, because it would be very expensive and disruptive to introduce another system. According to Paul Debenham, life sciences director of LGC, a leading British testing laboratory, "you could in theory introduce something different – based for example on the single-nucleotide polymorphisms, SNPs, that are causing a lot of interest in medical genetics – but that would not add much to the forensic scene" (Cookson, 2003, p. 9).

The emphasis now is on speeding up and miniaturizing DNA extraction and testing. Laboratories used to take a month or so to analyse and report on crime-scene samples. Now this can be done within five days. The ultimate goal is to have a box in the police car that could give a DNA profile within a few minutes and send it by a mobile communications link to the national database (Cookson, 2003).

Geneticists are also interested in the prospect of deducing clues to a suspect's physical appearance from a DNA sample. The first way of doing this is indirect. Different population groups have different patterns of genetic variation in STRs and SNPs, and by analysing the pattern of a DNA sample it may be possible to predict the "racial" origin of a criminal. For instance, the UK Forensic Science Service is beginning to offer an "ethnic inference service". This test can give the police a series of probabilities, which may provide useful guidance if they have no other information. The probability that the DNA comes from someone in a particular ethnic group may be as high as 90 per cent. The second way is a direct method that looks at genes coding for proteins that influence physical features. At present this can be done reliably only for hair colour, analysing differences in the MC1R gene that encodes a molecular switch involved in hair pigmentation. Forensic tests for eye and skin colour are being developed and many geneticists believe it will be possible to predict facial structure from DNA samples. But Jeffreys cautions against expecting too much, because we understand very little about genetic control of skin colour, let alone facial characteristics, which are extremely complex (Cookson, 2003).

Sampling DNA from corpses, with the prior consent of the families concerned, is considered a promising area for research in terms not only of genealogy and paternity tests but also of the genetics of diseases. In the United States, this kind of service is offered by undertakers associated with specialist laboratories. It is also being extended to European countries. For instance, in Spain, the company Intur, in collaboration with General Lab, is developing a DNA bank with samples from corpses. The first tests have been carried out by Serveis Funeraris of Barcelona, a company whose shareholders are Intur (49 per cent) and the City Hall (51 per cent). This service is offered to the company's clients in addition to its conventional work (Peña, 2004).

The sampling process, which had been requested by six people in Barcelona by mid-May 2004, needs an initial request from a relative, who signs a contract with Serveis Funeraris to do with the sampling procedure and another contract with General Lab regarding the extraction of DNA and its storage in the database. The undertaker's role is just to establish the contact between its clients and the laboratory. In the case of General Lab, the storage of the sample is initially foreseen for five years, but the family can extend the period. The cost of sampling and DNA extraction is €120–130; if the five-year storage is included, the cost rises to €250 (Peña, 2004). According to Eduardo Vital, executive officer of Serveis Funeraris, the extracted material may be of major medical interest, because it will help to identify the DNA of previous relatives of the person who has passed away and thus contribute to a better knowledge of the diseases affecting the family members as well as the risks for descendants (Peña, 2004).

One factor that highlights the importance of DNA extraction and conservation is the increasing trend of incinerating the corpses; in Barcelona, the incineration rate is 30 per cent and is tending to rise. Consequently, if one wishes to conserve the DNA of a dead person, this should be done before the body is incinerated. The current law in Spain requires the presence of a witness during the sampling process, so as to avoid any misconduct. However, the sampling of cells and DNA extraction can be psychologically very disturbing for the witness, especially when tissues have to be removed for the extraction of cells best suited for isolating their DNA (Peña, 2004).

3

Regulatory issues

Drug approval in the United States

According to the Tufts Center for the Study of Drug Development at Tufts University, the period for approval of new bio-pharmaceuticals in the United States decreased by 21 per cent between 1982 and 2002, from 24.9 months to 19.7 months. At the same time, the average time needed for the clinical trials of new products increased by 137 per cent, from 31.2 months to 74.0 months. The overall conclusion is that the combined average of the periods for development and approval of new drugs by the US Food and Drug Administration (FDA) was three years longer in 2000–2002 than during the 1980s: 7.8 years compared with 4.7 years. Janice Reichert, the principal investigator behind the Tufts report, stresses that nowadays much more complex and innovative substances are being developed and tested, without any guarantee that they will become effective new drugs. Most of the biotechnology-derived products approved during the 1980s were proteins whose functions were well understood (e.g. the anti-haemophilic factor VIII, human insulin and growth hormone). The Tufts report indicates that 70 per cent of biotechnology-derived products approved between 2000 and 2002 were recombinant proteins. During that period, the average time for developing DNA-recombinant products was 19 per cent shorter than the time required to devise monoclonal antibodies, but 13 per cent longer than the time needed for compounds of biological origin. However, DNA-recombinant products were approved by the FDA more quickly than monoclonal antibodies (+72 per cent), although 75 per cent of these products had been submitted on a priority basis to the FDA.

Efficient risk management

It is worth mentioning the FDA's new strategic plan published in mid-August 2003 by Mark McClellan, the head of the FDA – a medical doctor and an economist. What he calls "efficient risk management" means that, given its inadequate budget, the FDA cannot do all that it is charged with doing; so it must focus on those things where it can have the greatest impact, and do them more efficiently than in the past (*The Economist*, 2003b).

A major challenge for the pharmaceutical industry is to increase the efficiency of the drug approval process, and so reduce development costs. The United States is the world's most innovative drug market, but its pace of new drug approvals has slowed. In 2002, the FDA allowed seven high-priority new drugs on to the market, after an average review period of 14 months, which was less than half as many approvals as in 1996 and taking more than two-thirds longer to approve. In large part this was owing to factors beyond the FDA's control, not least the failure of drug-makers to turn bright ideas or discoveries into products. But the head of the FDA sees inefficiencies in the approval process that he wants to remedy. The FDA has always simplified its drug-reviewing machinery by shifting most of its biotechnology-derived drug reviews into its main reviewing centre, hoping to gain economies of scale. It is trying to make better use of external expert resources – for instance, by working more closely with the National Cancer Institute to help improve its drug-reviewing process. It also plans to interact with drug companies earlier in the development process to identify what sort of tests they should be doing, and also to inform them of potential FDA objections sooner. The FDA wishes to rely on good statistical methods. Drug tests often require firm evidence that patients live or die, which is necessarily quite a long process. Yet there are other markers, such as biochemical changes, that are statistically highly correlated with success and indicate effectiveness far earlier. They have already been used to approve some anti-HIV/AIDS drugs, for instance. The FDA is keen to extend the use of such statistical markers to other kinds of drugs, such as those against cancer (*The Economist*, 2003b).

The FDA's new approach to efficient risk management will face the Agency's usual critics, of whom there are two main groups. Anything that speeds up and reduces the cost of drug development will please the first group, the drug companies. But it may displease consumers activists, who argue that the FDA's attempts to speed up the approval process have resulted in an inefficiently high staff turnover rate in the Agency and, more importantly, a rising number of drugs that are approved and later have to be withdrawn. It is true that the US public is angry about what firms are up to, particularly in regulated industries. Yet, in August 2003, almost a year into a job so controversial that it took nearly two years to fill, the head of the FDA was still well received by all sides of the industries he oversees. Although the FDA cannot directly deliver cheaper drugs because it does not regulate drug prices, in August 2003 the Agency eased the way to a market for generic drugs (*The Economist*, 2003b).

Another example of the FDA's approach to efficient risk management is dealing with its new responsibility for the security of food supplies, and in particular of imports, in the wake of the September 11 attacks. The Bioterrorism Act, which took effect on 12 December 2003, aims to monitor the distribution of food in the event of a contamination emergency related to terrorist activity. The FDA has ordered food producers to keep files of all their operations and to make electronic reports one to five days before their cargo arrives at US Customs. If the reports do not look right, shipments will be inspected by Customs agents, who can hold or even destroy them. Rather than check shipments at the border – which is very expensive – the FDA plans to develop a detailed database to provide a profile of imports that will allow it to target its inspections and to work more closely with the governments of countries of origin.

Companies stated that the process could further slow entry of their merchandise into the United States. Some argue that the FDA requirements violate the North American Free Trade Agreement (NAFTA), which aims to streamline trade among Canada, Mexico and the United States. The president of the Food Council of 220 Mexican exporters believed that the new regulations would hit food exports. Exporters in Canada and Mexico claimed that these rules – which require a 24-hour waiting period – would hamper their "just in time" shipments, whereby firms despatch orders to the United States a few hours after receiving them. Exporting costs will also rise, because companies will need agents to represent their interests in the United States if they do not already have distribution channels, and they will have to pay to warehouse merchandise retained by Customs. Mexican business people believe the new rules will deal another blow to already weak sales, reflected in lagging food exports to the United States, Mexico's main trading partner. In 2002, Mexico exported food worth US$3.86 billion, about 90 per cent of it to the United States, and Canada exported US$17.4 billion in food, 67 per cent to the United States. In the month of July 2003, Mexican food and beverage exports totalled US$351 million, down 5.7 per cent from the US$371 million exported in July 2002, according to the Central Bank (Moreno, 2003).

Drug approval in the European Union

In the European Union, pharmaceutical companies can still choose to file applications for new products with regulators in individual countries, but biotechnology-derived products have to go through the European Union regulator, the European Agency for the Evaluation of Medicinal Products (EMEA), which was set up 1995. In October 2003, the chief executive of Serono SA, the Swiss company that is third biggest in the bio-industry, stated that this centralized drugs approval process was too bureaucratic and took place "behind closed doors". He said that no one took responsibility on the approvals committees, which have 30 members (two from each country). Decisions were made in private sessions and appeals were heard by the same committee. His remarks were prompted by the EMEA's decision in April 2003 to reject Serono's Serostim drug – a growth hormone for AIDS-related wasting, which has been available in the United States since 1996. The EMEA stated that Serostim was the first AIDS drug application it had rejected (Dyer, 2003a). Serono's chief executive also claimed that the Agency's approvals process was slowing the introduction of new drugs and putting pressure on prices in Europe, thus damaging the industry's competitiveness (Dyer, 2003a). The Agency's response was that, from mid-2004, EMEA's approvals committees would be limited to one person from each EU country plus five scientific experts. Moreover, for a number of diseases, the Agency had set up separate advisory panels; and it declared that "we have made considerable progress in opening up the process".

4

The economics of pharmaceutical biotechnology and bio-industry

The global pharmaceutical market

According to a study by Frost & Sullivan Chemicals Group, the global pharmaceuticals industry is forecast to grow at a brisk rate of 15 per cent a year over 2005 and 2006 (Adhikari, 2004). An even more optimistic projection, based on regulatory approval for a clutch of critical drugs (Genentech, Inc.'s Avastin for colorectal cancer; Imclone/Bristol-Myers Squibb's Erbitux, a monoclonal antibody against advanced colon cancer; Chiron Corp.'s Proleukin to treat HIV/AIDS), forecasts global revenue growth of over 25 per cent a year. In 2003 alone, 50 biopharmaceutical drugs from public biotechnology companies were in phase-3 clinical trials in Europe, of which 10–15 were likely to reach the consumer market by 2006–2007.

According to Ernst & Young, the biopharmaceuticals market had revenues of US$41.3 billion in 2002, out of a total US$350 billion for the whole drug market. Assuming a growth rate of over 25 per cent per year, global revenues will reach approximately US$95 billion by early 2007 (Adhikari, 2004). In 2002, the leading biopharmaceutical in revenue terms was Procrit (recombinant Protein-EPO), which generated US$4.27 billion for OrthoBiotech/Johnson & Johnson. Other leading products were Amgen's Epogen, Neupogen and Enbrel, Centocor's Remicade, Genentech's Rituxan, Biogen's Avonex, Eli Lilly & Co.'s Humulin and Humalog, GlaxoSmithKline's Combivir, and Berlex Laboratories' Beta-seron (Adhikari, 2004).

Consolidation in the pharmaceutical industry

In 2004–2005, concentration in the global pharmaceutical industry was still in progress. In 1996, the Swiss companies Ciba-Geigy AG and Sandoz AG merged into Novartis AG. This was followed in 1999 by the mergers of the French firm Rhône-Poulenc and German pharmaceutical group Hoechst AG, which became Aventis, and of Astra (Sweden) and Zeneca (United Kingdom). At the end of 1999, Pfizer, Inc. bought Warner-Lambert and Pharmacia, becoming the world's leading pharmaceutical corporation.

Although Novartis AG's headquarters are in Basel, Switzerland, 50 per cent of its annual turnover is made in the North American market. Novartis employs about 79,000 people and is present in 140 countries. In contrast with many pharmaceutical groups, Novartis has never hidden the fact that it was in continuous external growth. The merger between Ciba-Geigy and Sandoz, two very wealthy and profitable groups, was hailed as the "biggest operation ever tried in the global industry" – the new group being worth more than €20 billion at that time. The merger was considered a model for similar mergers and acquisitions during the 1990s. Currently Novartis AG's capitalization on the stock market is about €92 billion and its research portfolio is amongst the most successful in the world (Mamou, 2004c).

Novartis's policy of acquiring companies and laboratories is based on the fact that the development of new molecules and drugs is increasingly costly and research output is not very promising in terms of really innovative medicines. Consequently, instead of investing colossal funds in high-risk R&D, it is better to purchase competitors and their market share. This policy resulted in the top 10 pharmaceutical groups controlling over 60 per cent of the market in 2004 (Mamou, 2004c).

In 2003, Novartis was authorized to market five new drugs: Certican and Myfortic for transplant operations, Stalevo for the treatment of Parkinson's disease, Xolair in the United States for allergic asthma, and Prexige in the United Kingdom. Novartis's top 10 drugs in terms of sales values in 2003 were: Diovan (anti-hypertension), US$2,425 million; Gleevec (anti-cancer), US$1,128 million; Neoral (anti-transplant reject), US$1,020 million; Lamisil (anti-infectious), US$978 million; Zometa (anti-cancer), US$892 million; Lotrel (anti-hypertension), US$777 million; Lescol (anti-cholesterol), US$734 million; Sandostatin, US$695 million; Voltaren (anti-arthritis), US$599 million; and Cibacen (anti-hypertension), US$433 million (Mamou, 2004c).

Daniel Vasella, the chief executive officer of Novartis AG, who has been promoting his group's acquisition policy, stated that pharmaceutical groups will have to adapt to the increasing demand for efficiency and quality from industrialized countries' consumers; the winners will be not necessarily the biggest groups but the ones that respond fastest. In addition, the activity of these groups will depend, to some extent, on their attitude toward contributing to solving health issues in the developing countries (Mamou, 2004c).

Pfizer remains the world leader, with an annual turnover of US$38.2 billion and profits of US$4.7 billion at the end of 2003. GlaxoSmithKline is second, with US$32.2 billion and US$6.7 billion, respectively; Sanofi-Aventis is third, with US$25.8 billion and US$4.6 billion (estimates); Novartis is fourth, with US$21.1 billion and US$4.25 billion; Merck & Co., Inc., is fifth, with US$19.1 billion and US$5.8 billion; and AstraZeneca is sixth, with US$16.0 billion and US$1.9 billion (Orange, 2004).

Pfizer strengthened its lead as the world's largest drug-maker in April 2003 when it acquired rival Pharmacia Corp. for about US$57 billion. Excluding one-time charges and asset sale gains, Pfizer's 2004 first-quarter earnings rose 27 per cent to 52 cents per share, compared with 41 cents in the same period in 2003. Sales rose 47 per cent to US$12.5 billion in the quarter, owing to Pfizer's acquisition of Pharmacia, strong product sales and a weaker US dollar. The group's aggressive acquisition strategy has given it synergies, allowing it to cut costs while building a diversified product portfolio that can shield it against surprises. It also brought it the most lucrative drug in the world, Lipitor (to counter cholesterol accumulation). Lipitor led Pfizer's strong sales growth, with a 19 per cent increase to US$2.5 billion in the first quarter of 2004, which saw Pfizer and Lipitor dominate the anti-cholesterol drug market. A two-year study by US rival Bristol-Myers Squibb showed that Lipitor was better at lowering cholesterol and preventing heart disease than Bristol's Pravachol. Another study showed that Pfizer's experimental treatment torcetrapib raised so-called "good-cholesterol", or HDL (high-density lipoproteins), representing another potential significant breakthrough treatment. Pfizer obtained torcetrapib in its US$1.3 billion purchase of Esperion Therapeutics in the first quarter of 2004. However, charges from that and other acquisitions, including pharmacy and veterinary businesses, affected Pfizer's first-quarter 2004 net profits (Bowe, 2004). Including US$1.5 billion non-cash charges and other items, Pfizer's net profits fell 50 per cent to US$2.3 billion, or 30 cents per share, from US$4.7 billion, or 76 cents per share, in 2003. However, Pfizer's first-quarter 2003 profits were boosted by its US$2.2 billion gain on the disposal of the Adams confectionery business and Schick-Wilkinson Sword, the shaving product division (Bowe, 2004).

On 3 April 2000, Pharmacia & Upjohn and Monsanto Co. merged, giving rise to a new company called Pharmacia Corp. less than four months after the initial announcement of the merger. Pharmacia Corp. became one of the top pharmaceutical corporations in the world, with sales of about US$11.3 billion. With the agrochemical and seed business of Monsanto, Pharmacia's annual turnover amounted to US$17.0 billion, and the group employed 60,000 people worldwide (Lorelle, 2000).

In 1999, Pharmacia & Upjohn's net profit was US$803 million, annual turnover having increased by 7.3 per cent to US$7.25 billion. Monsanto Co. had been trying to find a partner in pharmacy because of the number of potential drugs in the pipeline and the fact that it did not have the infrastructure needed to commercialize all those products. Although Monsanto's pharmaceutical division, Searle, accounted for only half the agricultural division's activities, it made twice as much profit. Before the merger, Monsanto sold its nutrition division (which included the sweetener brands Nutrasweet, Equals and Canderel) to Tabletop Acquisition for US$570 million in early 2000. Monsanto also had to pay US$81 million to Delta & Pine Land (D&PL) Corporation, after deciding, in December 1999, to break the merger agreement between the two corporations concluded in May 1998 (Lorelle, 1999b).

The new company, Pharmacia Corp., had the third-biggest sales infrastructure in the United States. This was devoted to selling Monsanto's "star" drug Celebrex, which, after its commercialization in 1999, became the first prescribed medicine against arthritis, with sales of more than US$1 billion in less than 10 months; and Xalatan, which was the first drug against glaucoma, with sales of US$507 million in 1999. Pharmacia & Upjohn was also one of the world-leading producers of Genotonorm (a human growth hormone synthesized via genetic engineering) and was commercializing Zyvox, the first of a new class of antibiotics – the oxazolidinones (Lorelle, 1999b, 2000).

One-third of Pharmacia Corp.'s sales were in plant protection and the agricultural sector, considered by investors to be more risky than pharmacy. To reassure Pharmacia shareholders, the new company president, Fred Hassan (who was the chief executive officer of Pharmacia & Upjohn), explained in February 2000 to the biggest US investors that he intended to focus on the highly profitable pharmaceutical sector and to convert the agricultural branch into a subsidiary under the name of Monsanto Co. The new subsidiary was to become a distinct legal entity with an autonomous board of trustees and its own public shares (Lorelle, 2000).

On 30 October 2000, Pharmacia Corp. met Wall Street expectations for its third quarter. Adjusted after-tax profits were US$427 million and earnings per share 33 cents, up 57 per cent year-on-year. On the pharmaceuticals side, the group saw sales rise by a fifth, to US$3.28 billion in the quarter. The strongest progress came in the domestic market, up 34 per cent, buoyed by the continued growth of Celebrex sales, which totalled

US$687 million in the quarter and US$1.8 billion for the first nine months. In the United States, its third-quarter sales were US$584 million. In the same period, sales for the rival Vioxx product, made by Merck & Co., were US$615 million worldwide and US$495 million in the United States (Tait, 2000).

Meanwhile, Monsanto Co. – 85 per cent owned by Pharmacia, following an initial public offering for the other 15 per cent – had made a net loss of US$66 million, almost half the US$127 million deficit it made a year previously, on a 2 per cent sales gain to US$1 billion in the seasonally slow quarter. The net loss before restructuring charges was a smaller US$45 million, which was US$38 million better than a year previously. The company said that reduced R&D spending, coupled with tighter cost controls, had cut expenditures by about US$63 million, or 12 per cent. Monsanto was able to raise US$700 million after a first introduction on the stock exchange, in order to stimulate the market for genetically modified crops and products; the overall operation was less successful than expected, however, shares being traded at US$20 instead of the anticipated US$21–24 (Tait, 2000).

Japan, with its quasi-protectionist regulatory regime, has for years been considered to have a pharmaceutical market that is extremely difficult to penetrate. As a result, Japanese companies flourished at home but never felt the need to expand internationally. However, with its ageing population and budgetary difficulties, Japan can no longer afford to pay inflated prices for medicines from protected domestic corporations. Drug prices have in fact been slashed and reimbursements skewed toward more innovative drugs, the bulk of which are produced by foreign companies. In addition, rules on drug approvals have begun to converge internationally, making it much easier for foreign firms to launch drugs on their own in Japan. AstraZeneca is the fastest-growing drug company in Japan, Pfizer has the largest number of sales representatives, Merck & Co. took over its long-standing partner Banyu, and Roche AG bought Chugai (Pilling, 2004).

Consequently, the more far-sighted and innovative Japanese drug companies realized that, if their local market was being invaded, their best chance of survival was to expand abroad. For instance, Fujisawa's best-selling drug – a transplant medicine called Prograf – reaps almost all of its profits in the United States and Europe. Prograf is a niche drug that can be marketed with a very small sales team to specialist transplant surgeons. But Japanese companies lack the scale to market a blockbuster drug to general practitioners (Pilling, 2004).

Another driving force toward concentration or mergers is the huge of cost of developing new drugs. Nearly all Japanese drug groups are too small to pay for the cripplingly expensive development of drugs beyond the early discovery stages. Pooling R&D budgets is the obvious solution, although, given the slow pace of change in the past, it is unlikely that there will be a rush of pharmaceutical mergers. However, the news at the end of February 2004 that Yamanouchi was buying Fujisawa to create Japan's biggest drug group after Takeda created considerable interest (Pilling, 2004).

This merger has more to do with Japan's US$50 billion drugs market – bigger than those of Germany, France and the United Kingdom combined – than with the potential of Japanese drug companies seriously to challenge the industry's heavyweights. Ranked seventeenth in the world, Yamanouchi–Fujisawa's annual sales of about US$7.5 billion will be less than a fifth of Pfizer's. Its combined R&D budget of US$1.3 billion is respectable, but small compared with the US$3–5 billion that the industry leaders spend on R&D every year (Pilling, 2004). Such is the growing feeling that Japanese companies need to join forces to create national champions that Yamanouchi could merge again. Depending on the partner, that could raise it among the world's top ten or even top five; if such a merger happened, it could transform Japan into a player in the global pharmaceutical market (Pilling, 2004).

On 25 April 2004, the merger occurred between the French company Sanofi and the French-German group Aventis. Through the offer made by Sanofi towards creating an attractive new entity, Aventis was valued at €55.3 billion, instead of €48.5 billion in an earlier offer; Aventis's share price was therefore worth €68.93, and its stakeholders were able to exchange five Sanofi shares plus €120 for six Aventis shares. Although Aventis's chief executive officer is leaving the new group, which will be headed by Sanofi's CEO, the latter made a commitment not to lay off anyone and not to transfer research to the United States to any great extent (Orange, 2004).

Aventis had been created in December 1999 as the result of a merger between the French company Rhône-Poulenc and the German pharmaceutical firm Hoechst AG. It had 75,570 employees at the end of 2003 and marketed several well-known drugs, such as Allegra (against allergies), Lovenox (anti-thrombotic), Lantus (against diabetes), Taxotere (anti-cancer) and Delix (anti-hypertension). Sanofi had been born in December 1998 as the result of a merger between the two French pharmaceutical companies Sanofi and Synthélabo. Sanofi had 32,440 employees at the end of 2003. Its highest-selling drugs were anti-thrombotic and hypertension medicines, as well as an anti-insomnia medicine (Orange, 2004). Sanofi's research was mainly concentrated in France; through the merger with Aventis, it has acquired an international dimension and quite a strong marketing capacity in the US market. The French government stated its great appreciation that the merger had created the biggest

European pharmaceutical group and the world's third biggest, and that this had avoided the involvement of a non-French company such as the Swiss–US Novartis, which had shown an interest in acquiring Aventis (Mamou, 2004d).

Will the merger between Sanofi and Aventis guarantee France a key position in the global pharmaceutical market? It is not that certain, because the success or even survival of a pharmaceutical group, whatever its nationality, depends on its ability to make an annual profit per share that is close to the average of most comparable enterprises (i.e. 20–25 per cent for the best groups), while at the same time devoting 18–20 per cent of its turnover to research. Both ratios depend on the price of drugs. In France, where this price is fixed by the government, it is increasingly difficult for pharmaceutical companies to meet the needs of both stakeholders and researchers. That is why nowadays most pharmaceutical groups wish to make 50 per cent of their annual turnover on the US market, where the prices of drugs are not fixed (Mamou, 2004d).

This major difference between the United States and the rest of the world affects investments. The pharmaceutical groups do not generally publicize the magnitude of their investments, but they are usually made where expected profits are higher, which most of the time means the United States. Consequently the R&D centres of the new group Sanofi-Aventis are not threatened, but they run the risk of being outpaced by US groups or companies. A recent study by Rexecode shows that the US share in global drug production increased by 5.3 per cent between 1986 and 2000, whereas Germany's share decreased by 3.1 per cent and France's by 1.7 per cent (Mamou, 2004d).

To sum up, mergers in the pharmaceutical industry can be explained by the increasing need for new drugs for an ageing population and corollary diseases. This demographic aspect is a main source of the profits of the pharmaceutical industry, which is the most profitable in the world with net profits equal to 15–20 per cent of annual turnover. However, the relentless rise in health expenditure in all countries – higher than the rise in gross domestic product – resulting from the health-care needs of an ageing population prevents most governments (with the exception of the United States) from allowing pharmaceutical groups the freedom to price their products. Thus, although the pharmaceutical industry has to invest increasing funds in R&D – the development of a drug costs US$500–800 million, compared with US$54 million in 1979 – health-care budget deficits lead governments to reduce the price of pharmaceuticals (Mamou, 2004b).

Another fear which leads to mergers in the pharmaceutical industry relates to the marketing of generic drugs. Since the mid-1990s, pharmaceutical companies have successfully chosen the "blockbuster" strategy, i.e.

the production of worldwide pharmaceuticals that can generate an annual turnover of more than US$1 billion. Such blockbuster drugs attract the producers of generic drugs who can market a copy of a drug before its patent expires. But the race between the pharmaceutical laboratories and generic drug companies is meaningless if there is not a steady flow of new compounds to replace those that fall in the public domain. There has in fact been a slowdown in the production of novel medicines. For instance, in the United States, 17 drugs were approved for commercialization in 2002, compared with more than 50 in 1982. Consequently one may wonder whether this slower pace of innovation will last for an industry that is producing less and facing increasing costs. The obvious response, as a counter-risk measure, is to acquire the promising pharmaceuticals in a competitor's portfolio (Mamou, 2004b).

Bio-manufacturing capacity

According to Frost & Sullivan, the total capacity demanded in Europe and America is likely to grow from 1.1 million litres (70 per cent North America and 30 per cent Europe) in 2003 to 3.0 million litres in 2010. Total capacity supplied across the global biopharmaceutical industry is set to grow from 2.0 million litres in 2003 to 3.5 million litres in 2010. Simultaneously, capacity utilization rates are set to increase from 55 per cent in 2003 to 85 per cent in 2010. This deficit is expected to affect industry structure, prices, revenues, R&D initiatives, product availability and manufacturing operations. According to Frost & Sullivan, it is also poised to trigger support for mergers and acquisitions (Adhikari, 2004).

Increased expansion activity is projected to push manufacturing capacity to over 3 million litres in 2006. In Europe alone, nearly 400,000 litres of manufacturing capacities are likely to be added by contract manufacturing organizations such as Boehringer-Ingelheim in Germany, Diosynth in the Netherlands and Lonza in Switzerland and the Czech Republic over the six-year period 2005–2010. Demand for manufacturing capacities is poised to climb following the approval of several promising new drugs. In particular, the 350-plus drugs undergoing clinical trials are likely to prompt more expansions while generating a sizeable demand for capacity. Already, demand for manufacturing capacities has shown significant growth owing to the approval of innovative drugs for anaemia, rheumatoid arthritis and other diseases. Ongoing trials of critical drugs against cancer, HIV/AIDS, diabetes and cardiac disease are also expected to boost capacity demand (Adhikari, 2004).

At the same time, process yield improvements and pioneering expression systems for large-scale manufacturing of biopharmaceuticals could significantly alter the capacity gap. For instance, transgenic or recombinant DNA technology has the ability to deliver large manufacturing capacities at much lower production costs than current expression systems (Adhikari, 2004).

The strategies of biotechnology companies and pharmaceutical groups

There are two main categories of biotechnology companies: those that sell various services to the pharmaceutical groups, such as chemical and biochemical trials and tests; and those that carry out medical research. The companies in the first category have often developed technological platforms that enable the large pharmaceutical groups to streamline some stages of their research, e.g. the screening of compounds that may give rise to a medicine. This does not mean that the know-how concerning the screening of molecules will be completely abandoned – the pharmaceutical group having an interest in controlling its subcontractor – but the gain in productivity cannot be underestimated. The companies in the second category specialize in research and development of drugs; because they generally have limited resources, they try to conclude agreements with the biggest pharmaceutical groups (Mamou, 2004a).

In 2001 and 2002, the value of biotechnology companies' shares on the stock market plummeted and equity was scarce. In 2003 and 2004, confidence was back but investments in that sector were very selective. The lower number of companies becoming public (4 in 2002 and 7 in 2003, compared with 58 in 2001) and the scarcity of private investments in a sector that is greedy (US$16.4 billion in 2003 compared with US$32.0 billion in 2000) have led to an in-depth restructuring of the biotechnology sector. Small companies and start-ups have not completely disappeared and many of them remain a driving force in terms of the creativity of the sector, but the gap between them and the bigger biotechnology companies or the large pharmaceutical companies has widened. Thus, 60 per cent of the 1,500 biotechnology companies in the United States have between 1 and 50 employees, according to a study carried out by the US Department of Trade, and they are having to tighten their belt. At the other end of the spectrum, 1.9 per cent of the companies have more than 15,000 employees and earn revenues that are comparable with those of conventional laboratories (Mamou, 2004e).

The trend towards partnerships and mergers between big pharmaceutical groups and small biotechnology companies is partly explained by the crisis of productivity in conventional pharmaceutical research as well as by the steady erosion of patents by the generic drug industry. According to a study carried out by McKinsey in 2000, the number of new drugs marketed by each big pharmaceutical group fell from 12.3 over the period 1991–1995 to 7.2 over the period 1996–2000 (Mamou, 2004e).

Many biotechnology start-ups try to conclude financial agreements with pharmaceutical groups or to merge in order to better resist. Thus the British company Oxford GlycoSciences (OGS) was supposed to merge with Cambridge Antibody Technology, then with Celltech (which offered €147.7 million). Neither merger occurred, but the company's CEO was willing to accept the best offer in cash or shares (Lorelle and Ducourtieux, 2003). Other prominent examples of this trend are the takeovers by Novartis Pharma of Idenix Pharmaceuticals, by Hoffmann-La Roche of IGEN International, and by Chiron Corp. of PowderJect Pharmaceuticals, the acquisitions by Johnson & Johnson of 3-Dimensional Pharmaceuticals and by Union Chimique Belge of Celltech (see chapter 1) (Adhikari, 2004).

In fact, a key factor in the dynamism of the US biotechnology sector is the relentless movement toward partnerships, or even acquisitions, between the big pharmaceutical companies and biotechnology start-ups. The big pharmaceutical companies are convinced that their future lies in the 400 molecules being developed by the biotechnology start-ups, among which they may find a few blockbuster drugs that will earn annual sales above US$1 billion. The following examples illustrate this trend of partnerships and acquisitions:

- Amgen announced in mid-October 2000 the purchase of Kinetix Pharmaceuticals for US$170 million in shares, with a view to accelerating the development of its oral medicines;
- Idec Pharmaceutical acquired Biogen, Inc. in 2003 for US$6.9 billion;
- Pfizer, Inc. and Johnson & Johnson purchased Esperion Therapeutics and Scios, respectively, for US$1.3 billion and US$2.4 billion.

Merck & Co., the world's fifth-biggest pharmaceutical group, has for several years avoided any spectacular merger or aggressive purchase. Instead, the US company has a strong policy of partnerships with biotechnology firms and other pharmaceutical groups. About 40 such agreements were concluded in 2003, compared with only a dozen in 1999; another 80 were being examined. For instance, partnerships have been set up with Genpath for cancer, Amrad for respiratory diseases, Neurogen for pain-killers and Actelion (Switzerland) for cardio-vascular diseases. By the end of 2003, Merck & Co. acknowledged a number of setbacks. On 12 November 2003, it had to abandon phase-3 clinical trials meant to establish the effectiveness of an anti-depressant drug; on 20 November 2003, another drug against diabetes, also in phase-3 trials, had to be discarded, owing to the appearance of a malignant tumour in mice. In one week, therefore, two drugs with a high commercial potential disappeared from Merck's portfolio. In addition, Merck's anti-cholesterol blockbuster drug Zocor will lose its patent protection in 2006 (Mamou, 2003b).

On 9 December 2003, Merck's CEO, Raymond Gilmartin, tried to regain the trust of some 300 financial analysts and investors by playing down his company's recent setbacks and by showing that there were future blockbuster drugs in the pipeline. Thus, in early November 2003, an innovative anti-cholesterol drug (ezetimibe/simvastatin) had been submitted to the US Food and Drug Administration. By early 2004, Merck & Co. intended to request authorization for Proquad – a drug against measles, mumps, rubella and chickenpox. Between 2005 and 2006, another four products could be commercialized: a vaccine against rotaviruses, which affect millions of children; a drug against shingles; a drug against the human papilloma virus, which is a major sexually transmitted disease; and a probiodrug against diabetes (Mamou, 2003b). The company also set up a drastic cost-reduction plan. In October 2003, 4,400 lay-offs were announced for 2004 with a view to saving US$250–300 million as of 2005 (Mamou, 2003b).

On 26 March 2003, the Swiss company Novartis Pharma announced a 51 per cent stake (for US$225 million) in the US biotechnology company Idenix Pharmaceuticals, with a US$357 million commercial agreement over a promising drug against hepatitis C. Idenix, which is also located in France, specializes in the development of anti-viral drugs against hepatitis B, hepatitis C and AIDS. Through the purchase of Idenix, Novartis is strengthening its position in the treatment of hepatitis C; it may also be tempted to acquire another Swiss company, Roche (in which it has 32.7 per cent of voting rights), which has focused on the struggle against hepatitis C. Novartis has not followed the usual behaviour adopted by most of the big pharmaceutical groups, i.e. not to purchase start-ups but just to use them as suppliers of promising molecules in order to complete their range of marketable products (Lorelle and Ducourtieux, 2003).

It is true that the big pharmaceutical groups are mainly interested in blockbuster drugs, so as to be able to recoup their heavy investments in R&D and in marketing and sales operations. For example, Viagra, introduced in 1998, is one of Pfizer's best-selling drugs, with revenues of US$1.7 billion in 2002. There is great interest in finding new drugs for the fast-growing impotence market; the major drugs companies are in desperate need of new treatments to bring to the market because their patents on existing products are expiring. The current competition among drugs corporations and the future challenges in this regard are strikingly illustrated by the fierce marketing battle between four pharmaceutical groups over treatments for impotence. In a trial of Viagra, Eli Lilly's Cialis and Levitra (jointly marketed by GlaxoSmithKline and Bayer AG), the results of which were announced in November 2003 at a medical conference in Istanbul, 45 per cent of the 150 men involved preferred Cialis, whereas 30 per cent voted for Levitra. The 150 subjects took all three tablets at least six times before giving their preferences. Patients stated they had favoured Cialis because of the longer duration of erection and because the new drugs had fewer side-effects than Viagra (Dyer, 2003b). Pfizer claimed that the research, conducted by Hartmut Porst, associate professor at the University of Bonn, had numerous flaws, including the fact that both physician and patient knew which drugs were being taken, in contrast to the "blind" procedures used for rigorous scientific comparisons. The company announced that another study published in November 2003 at the European Society of Sexual Medicine meeting in Istanbul showed Viagra's benefits (Dyer, 2003b).

Eli Lilly, the US maker of the Prozac anti-depressant, announced on 5 January 2004 that its profit, excluding certain items, would be as low as 65 cents a share, because of the costs of introducing new drugs. The forecast excluded a "substantial" expense relating to the pending acquisition of Applied Molecular Evolution. The average first-quarter profit estimate by analysts surveyed by Thomson Financial was 67 cents. Profit for the year, excluding certain items, would be US$2.80–2.85 a share, held back by marketing expenses as Eli Lilly rolls out six new products. The company won FDA approval for its Symbyax bipolar treatment at the end of 2003 and for the impotence drug Cialis in November 2003. According to analysts, the company's R&D efforts over the past few years came to fruition in 2004 and the full commercial benefit was expected in 2005 (*International Herald Tribune*, 6 January 2004, p. 12).

The Indianapolis-based pharmaceutical company, whose shares fell 15 per cent and closed at US$70.19 on the New York stock exchange in early January 2004, was counting on new medicines such as Cialis, the osteoporosis treatment Forteo, and Strattera for attention-deficit disorder to raise profits in 2004 as competition eroded Prozac's sales and growth slowed for Zyprexa, a schizophrenia treatment. In 2004, Eli Lilly anticipated having 11 major growth products, including six new products that were launched either in 2003 or in 2004. Spending on R&D in 2004 would rise in the mid-teens percentage range, compared with the 5 per cent growth that had been projected for 2003. The company reported that it had hired thousands of new sales representatives to promote its new products (*International Herald Tribune*, 6 January 2004, p. 12).

Aventis's Taxotere drug had been approved to treat lung cancer in the United States and Europe and gastric, ovarian, head and neck cancers in Japan. Taxotere belongs to a class of chemotherapy drugs known as taxanes, and the Franco-German pharmaceutical group wanted to have this drug recognized as the cornerstone of treatment regimens in a number of cancers. At the end of 2003, Aventis published the results from a study involving 1,500 women that showed that women with breast cancer who had the Taxotere-based treatment after surgery had a 30 per cent reduction in the risk of death after 55 months, compared with one of the standard regimens. These results were central to the group's application to US and European regulators in 2004 for the use of Taxotere in early-stage, operable breast cancer. At present, the treatment has been approved only for advanced or metastatic types of breast cancer after failure of chemotherapy (Dyer, 2003c).

This new therapy would help increase sales of Taxotere from US$1 billion in the first nine months of 2003 to US$2.6 billion in 2007. Chemotherapy drugs are facing competition from a new generation of targeted cancer treatments that have fewer side-effects, many of which exceeded sales expectations in 2001–2002. However, Aventis believed these new drugs were likely to be used in combination with chemotherapy (Dyer, 2003c).

The comparative advantages of biotechnology companies

For young pharmaceutical groups, in contrast to the big corporations, a market of only US$70 million is considered very important. On the other hand, the big corporations have not always kept the many gifted researchers who were attracted by the major changes occurring in cell and molecular biology; these are found nowadays in small companies, where they pursue their research and try to contribute to drug discovery. In addition, these small companies can find windows of opportunity in looking for treatments for diseases that are rather neglected by the big groups. In a way, they are trying to repeat the experience of those biotechnology companies that, in the late 1980s, were very successful through developing treatments that did not yet exist. This was the case of Amgen and Genentech, which developed recombinant erythropoietin (EPO) and human growth hormone, respectively. Nowadays, every investor dreams about having an Amgen in their portfolio, and enterprises dream of becoming such a prototype (*The Economist*, 2003a).

Genentech is a very good example in this respect. Founded in 1976 and located in the San Francisco bay area, it had about 6,500 employees in 2004 and an annual turnover in 2003 of US$3.3 billion. It is marketing 13 drugs and invested US$720 million in research in 2003. It was expecting to recruit 1,500 people in 2004 and to become a leader in cancer research by 2010. Genentech has specialized in anti-cancer monoclonal antibodies, to which was added the anti-colon cancer drug Avastin in May 2004. Another five products are to be commercialized. On another therapeutic front, disorders of the immune system, five drugs will be added in 2010 to the five already on the market. Finally, Genentech, which is a leading actor in the regulation of blood-vessel formation (angiogenesis), is working on three new products that will be available before 2010 (Mamou, 2004e).

Being a company whose activities range from research and development to commercialization, Genentech is no different from a big pharmaceutical group. It employs 800 sales representatives who are in charge of marketing its products to doctors, and it manages factories that account for 30 per cent of the world's capacity in biological products; the company also intends to distribute high dividends on its shares. Genentech is fully conscious of the risks associated with research and to mitigate them it is concluding licensing agreements and partnerships with other biotechnology companies and conventional laboratories. According to Hal Barron, Genentech's vice-president and director of medical research, more than US$500 million of the company's annual turnover derives from strategic agreements signed with other firms. Since 1999, more than 40 agreements have been concluded (Mamou, 2004e).

Like the big pharmaceutical groups, Genentech will be confronted with a decrease in its annual turnover owing to competition from generic drugs. Its first patents will expire in 2005. The only way out is to produce new drugs to compensate for the loss of old patents. At the same time, Genentech is fighting to defend its patented products, for example by trying to prevent a subsidiary of Novartis from copying its human growth hormone (Mamou, 2004e).

Many US biotechnology companies have oriented their research toward "orphan" diseases, whose number has been estimated at between 6,000 and 10,000, affecting 25 million people in the United States. Because the regulatory system is effective, the development of therapies for these diseases is faster and less costly than in the case of multifactorial diseases affecting massive populations, such as diabetes or atherosclerosis. Clinical trials can be shortened in the case of "orphan" diseases, regulatory aspects are simpler and the marketing of drugs in niche markets demands lower investments. Another important advantage is that patent duration is longer: commercial exclusivity cannot be less than seven years. According to the US Association of Pharmaceutical Manufacturers, more than 370 drugs were being developed by 144 US biotechnology companies and 234 products by 23 European enterprises in relation to rare diseases, as published in a report by Goldman Sachs on 14 July 2003 (*The Economist*, 2003a). In 2004, 1,300 drugs against "orphan" diseases were being tested in clinical trials and 250 had been authorized for marketing, including 11 in 2003 (Mamou, 2004e).

On 23 July 2003, the oldest French start-up, Transgène (now a subsidiary of BioMérieux), obtained the status of orphan drug for its treatment of T-cell skin lymphoma (the most widespread skin cancer after melanoma). This status will enable Transgène, which employs 165 people, including 138 researchers, to enjoy a 10-year commercial exclusivity so as to make its research more effective. Moreover, patients will have faster access to the new product (gamma-interferon) as of 2006 (Lorelle, 2003).

The European Agency for the Evaluation of Medicinal Products (EMEA) created the orphan drug status in 2000 and six out of eight orphan drugs marketed in Europe in 2003 were produced by start-ups. Meristem Therapeutics, a French company employing 84 people, also received, in July 2003, a favourable response concerning a substitute for gastric lipase, aimed at treating children suffering from cystic fibrosis. This is the first lipase produced through genetic engineering in genetically modified maize seeds, which could replace the lipase extracted from porcine pancreas in 2007 and thus avoid virus contamination (Lorelle, 2003).

Start-ups are also involved in finding treatments for diseases for which there are currently no effective therapies. Neovacs in Paris, in collaboration with Aventis Pasteur, has developed an anti-AIDS vaccine that boosts the production of antibodies against the virus after tritherapy has halted the multiplication of the virus; the vaccine is undergoing human trials. Faust Pharmaceuticals in Strasbourg (with 18 staff members), has obtained good results in the treatment of Alzheimer's disease; the same is true of Ethypharm for brain cancer, and of Carex in Strasbourg (with 25 staff) in the treatment of hypercholesterolemia (Lorelle, 2003).

Biotechnology start-ups can also take advantage of the relative lack of productivity of the big pharmaceutical groups as well as of the fact the big companies have recently abandoned large areas of medical research. Thus Aventis, whose two leading products were being threatened by their generic versions, and whose stock fell 3 per cent during the second quarter of 2003, aimed at multiplying its agreements for licensing and acquiring products, while at the same time screening strategic products from others that were less important for the growth of the group. In 2002, three non-strategic drugs were sold, and in 2003 three gastro-intestinal medicines sold to Axcan Pharma earned US$145 million. On 31 July 2003, Aventis purchased three drugs (in the pipeline) from the US company ImmunoGen for the treatment of blood, breast, lung and prostate cancers. Aventis had to pay US$12 million in advance, and another US$50 million by 2006, in addition to the rights on sales. The financial director of Aventis stated on 29 July 2003 that the group could not do everything and, while observing the worldwide expansion of biotechnologies, would target any relevant start-up, either to acquire it or to purchase promising drugs from it. Aventis's sales included 20 per cent of products from outside, compared with 40 per cent for its competitors (Griffith, 2003a).

On 9 December 2003, Aventis announced the purchase of its subsidiary Aventis Behring by the Australian group CSL Limited for US$925 million. Aventis Behring, based in Pennsylvania, is a leader in the market for therapeutic proteins (€1.06 billion turnover in 2002) and its drugs concern haemophilia, respiratory deficiency and several immunological and genetic disorders. CSL's annual turnover in 2002 amounted to €752 million (US$920 million) in biopharmacy. Aventis's focus on strategic products representing over 66 per cent of sales emphasized two targets: diabetes and cancer. "Reshaping Aventis" was a financial plan to be implemented as of 2004 in order to make savings of €500 million over the three-year period 2004–2006 (Mamou, 2003a).

In May 2003, Genentech made the surprise announcement that its experimental drug Avastin worked on colon cancer. In the same month, ImClone stated it would apply for approval of Erbitux, its monoclonal antibody against advanced colon cancer that initially was rejected by the FDA in December 2001. In September 2003, a flood of positive news about experimental biotechnology drugs led to fresh enthusiasm for investment in the sector. The Nasdaq Biotechnology Index (NBI) nearly doubled to 785.47 from its low earlier in 2003, although it was still at half its peak of about 1,600 in 2000, when hype over genomics was at its height. Mark Monane, biotechnology analyst at Needham, stated: "What we're seeing is the maturation of the biotech pipeline" (Griffith, 2003a, p. 19).

In June 2003, Biogen and Idec Pharmaceuticals merged to form the third-largest biotechnology company in the United States. Whereas most pharmaceutical groups' mergers were about cutting costs, Biogen and Idec's merger is about building pipeline, i.e. to develop more innovative molecules. The merger was friendly and the headquarters will remain in Biogen's home town of Cambridge, Massachusetts, rather than switch to San Diego. This means the company will stay close to Harvard, the Massachusetts Institute of Technology, the Whitehead Institute (a prominent research centre), Novartis AG's new research headquarters and scores of smaller biotechnology companies. Indeed, this location is part of the corporate strategy (Griffith, 2003b).

One doubt initially raised about the merger was that the two companies were too similar. Both had very successful drugs on the market: sales of Biogen's flagship product, Avonex (beta-1a interferon) for multiple sclerosis, reached US$1 billion in 2003; annual revenues from Idec's Rituxan (monoclonal antibody against cancer) were almost US$2 billion. But Biogen Idec's latest entrants – Amevive for psoriasis and Zevalin for non-Hodgkin's lymphoma – were unlikely to meet business needs; the pipeline looked thin, although a new treatment for multiple sclerosis, Antegren, was showing promise in human trials. All this presented one more argument for the merger; if a company did not have a full pipeline, it could always buy experimental drugs. And that is what Biogen Idec planned to do, using some of its US$1.5 billion cash. It will have to compete head-to-head with big pharmaceutical groups, themselves eager to acquire new drug candidates (Griffith, 2003b).

However, buying in molecules will not be enough. Biogen Idec must leverage its new resources to reinvigorate its in-house research as well, knowing that so many drugs fall apart in the clinic. For instance, in October 1999, Biogen realized that Antova, an experimental treatment for multiple sclerosis and diabetes, caused blood clots. This devastating announcement underlines that one has to place a lot of bets in the drug development process. Human clinical trials raise many complex issues, because it is not possible to unravel the thousands of chemical and biological reactions triggered by a drug in the human body. It is a field that has recently acquired a name of its own – "systems biology". Sometimes there are good surprises. In the case of the Biogen Idec merger, some analysts wondered what a cancer group, Idec Pharmaceuticals, could have in common with an immunology company, Biogen. It turned out that while investigating the anti-cancer drug Rituxan, for instance, clinicians were stunned to realize it appeared to clear up rheumatoid arthritis symptoms in patients suffering from both diseases (Griffith, 2003b).

Genentech and Xoma have tested their new anti-psoriasis drug with patients with psoriatic arthritis. Genta and its partner Aventis were seeking the FDA's approval for Genasense, their skin cancer treatment, on the strength of late-stage data released on 10 September 2003. Some scepticism persisted about Genasense, since about one-third of the patients in the study had not completed 12 months of treatment. But the company and many analysts believed the numbers were strong enough for approval. The market for Genasense was expected to be much larger, eventually, than skin cancer patients (Griffith, 2003a).

On 15 September 2003, the Medicines Company released a follow-up study on its heart treatment Angiomax, data the company expected would boost sales by 50 per cent. According to the company's chief executive, the "trial results were so positive that we expected Angiomax to become the gold standard treatment for angioplasty" (Griffith, 2003a, p. 19).

In September 2003, Aventis showed its confidence in medical biotechnology research by paying US$550 million for the rights to an early-stage cancer drug developed by the biotechnology firm Regeneron. The average collaboration for late-stage molecules soared from US$168 million in 1998 to US$207 million in 2000, according to the Biotechnology Industry Organization trade group (Griffith, 2003b). According to the Goldman Sachs' report issued on 14 July 2003, 67 per cent of the 234 drugs being developed in Europe were available to be licensed to big pharmaceutical groups, but some companies, such as Serono SA or Celltech, might not wish to look for partners, owing to their strong development and marketing capacities.

Building on their comparative advantages, biotechnology companies could therefore become future medium-weights in pharmacy. Cerep, founded in 1989, is one of the very few biotechnology companies in the world, and the only one in France, that combines both research and services. According to its chief executive officer, Thierry Jean, the company uses its regular income from the sales of services to fund its own drug research. In 2002, its annual turnover was €34.5 million; it sells 650 tests to about 300 clients in the pharmaceutical and biotechnological sectors. The company is trying to widen the range of its services in order to support its growth; for instance, when the services are closer to clinical trials on humans, the income is bigger. Thus, its 2003–2004 turnover was to be boosted by the clinical trials of a compound selected for Bristol-Myers Squibb that may be effective against inflammatory diseases (Mamou, 2004a).

On 15 January 2004, Cerep announced the purchase of the Swiss company Hespérion for €10.2 million, with a view to widening the scope of its activities towards clinical trials. Hespérion was connected to Actelion, a company created by former researchers from the Swiss pharmaceutical group Roche Pharma and which contributed to developing Tracleer, a treatment for lung arterial hypertension. The acquisition of Hespérion will enable Cerep to carry out clinical trials of potential drug compounds and to earn three to four times more money than from its usual chemical or biological research (Mamou, 2004a).

Urogène (55 staff), which specializes in bladder and prostate cancers, has purchased Chrysalon from Aventis. Urogène's clinical network, combined with Chrysalon's know-how in biology and chemistry, will enable the company to develop its own drugs. As for Hybrigenics, located in Paris, it purchased the Dutch firm Semaia at the end of April 2003, and aims to become a world leader in the discovery of oncological drugs. In January 2003, Meristem Therapeutics set up a commercial facility in the United States. Bioprotein, the world-leading company for the production of drugs from transgenic rabbits' milk, Urogène and the Economic Agency in Essonne opened a bureau in Boston in July 2003 (Lorelle, 2003).

Funding biotechnology research and development

Pre-seed investment

Pre-seed investment generally refers to investment in highly promising ideas at a very early stage – often to achieve proof of principle, firm up a business model and establish a management team. Pre-seed investment can also be made further downstream to advance a venture past a critical milestone or to develop a business model for an already proved idea. The role of pre-seed investment is to add value to the project so that it becomes attractive to investors, because venture capitalists are generally not interested in very early-stage investment. With pre-seed investment, the scientific and technical merits of the project are important, but these are not the only drivers in the decision to invest. Factors such as the size of the market, the attitude of inventors and the protection of intellectual property are also crucial (Dando and Devine, 2003).

Because of the limited capital available, there will be a focus on value-critical milestones and risk reduction. Consequently, the research programme will be targeted at reaching specific milestones that provide proof of principle or add value to the venture, and particular scientific interests may not be explored unless there is some firm commercial basis on which to do so. In this regard, the investor will be looking for committed inventors who are realistic about their role (Dando and Devine, 2003).

It is essential that the investment proposal includes credible market data and a commercialization model because this demonstrates that the people involved have an understanding of the market and a realistic valuation of the investment. For instance, there is little point in citing the worldwide sales figures of asthma drugs if the candidate drug will be effective in only a small proportion of patients suffering from this disease (Dando and Devine, 2003).

In Australia, UniSeed Pty Ltd is a US$20 million pre-seed venture capital investment fund established as a joint venture between UQ Holdings Pty Ltd and Melbourne University Private. UniSeed fills the gap between academic and commercial research, with investments of up to US$500,000 being made, with the possibility of follow-up funding. UniSeed has 10 active investments in the biotechnology/health sector (six in Melbourne, three in Brisbane and one in Sydney) with a further two new investments at term sheet stage. Most of these investments have been in ventures arising directly out of academic research, although one investment was in a medical device conceived and developed by a private investor. The following emerging companies are part of UniSeed's biotechnology and health portfolio (Dando and Devine, 2003):

- Adipogen Pty Ltd has identified a target for which antagonist compounds could be used as an effective anti-obesity therapy. A protein stimulates a number of pre-fat cells (adipocytes) and primes these for differentiation into mature fat cells. Inhibiting this protein may thus be an effective mechanism for controlling body fat mass. The project seeks to develop the technology through pre-clinical evaluation of lead compounds. Blocking compounds (mimetics) that mask the site of protein action will be developed and their efficacy tested in human adipocyte cultures and then in a suitable animal model.

- Cryptopharma Pty Ltd is a start-up dedicated to developing therapeutic treatments for anti-tumour and anti-inflammatory indications. Initial developments have centred on analogues of 2-methoxyestradiol, with a focus on asthma as the indication. A family of compounds, highly potent and selective inhibitors of smooth muscle proliferation, has been developed. Their anti-asthmatic activity has been demonstrated in an *in vivo* mouse model; they represented a totally novel therapeutic approach to treating asthma.
- Hepitope Pty Ltd is developing a novel DNA vaccine technology that promotes both humoral and cell-mediated immune responses. The system improves the efficiency of nucleic acid vaccines by targeting antigens to both major histocompatibility complex (MHC) I and MHC II pathways. The current focus is on developing vaccines against hepatitis B and C and HIV/AIDS.
- Pargenex Pty Ltd has designed and synthesized novel lead compounds as potent activators of newly elucidated drug targets (airway epithelial PAR2 receptors). These new lead compounds will provide the basis for novel drugs for treating inflammatory diseases of the airways such as asthma and chronic obstructive pulmonary disease.
- QRx Pharma Pty Ltd was formed in late 2002 with initial venture capital backing led by Innovation Capital (Australia and the United States), Nanyang Ventures (Australia), SpringRidge Ventures (the United States) and UniSeed (Australia). QRx Pharma has a highly experienced management team with experience in start-up and large companies in the discovery, development and regulatory approval of human therapeutics, particularly for the US market. QRx Pharma's initial objective is to commercialize proprietary technologies originating from the University of Queensland for the treatment of pain and the control of bleeding. The company had a development pipeline that included a product in phase-2 clinical trials and others in pre-clinical development and discovery.

Venture capital

The venture capital (VC) industry enjoyed an extraordinary period of easy money during the dotcom bubble. In 1999, the last full year of VC bullishness, buoyant stock markets made it absurdly easy to float companies, the most lucrative exit route for early-stage investors. For 1999, annual VC returns, comprising realized profits and the rise in the value of retained holdings, reached an unsustainable 146 per cent. By contrast, in the year to September 2003, the average American VC firm showed a "return" of −17.8 per cent. In 2004, there were signs of a revival and, although biotechnology investing was becoming fashionable again, VCs never lost their fondness for health-care starts-ups (*The Economist*, 2003a).

The lure of biotechnology and drug development has always proved irresistible, even though returns, so far, have been disappointing. However, as the drug industry has changed, so too has the way VC firms invest. Increasingly, they have to compete not just with each other but with funds that once specialized in only the biggest buy-out deals but are now willing to invest in start-ups as well. For instance, Kohlberg Kravis Roberts, which is famous thanks to its mammoth leveraged buy-out of RJR Nabisco (a food conglomerate) in the 1980s, made its first biotechnology investment in March 2004 in a US company called Jazz Pharmaceuticals, which is developing drugs for psychiatric diseases. That represented a major departure from its usual strategy of investing in more mature companies (*The Economist*, 2004a).

VC investments in biotechnology have become bigger in recent years. For instance, in 1997 the typical investment was less than US$10 million. Nowadays, funding rounds of US$20–25 million are more common. Drug development costs have risen steadily, but so have the rewards. These are good reasons why biotechnology deals might make long-term sense. The new-product pipelines of the world's biggest drug companies have recently been running dry. They are increasingly relying on much smaller start-up and VC-backed firms to provide them with a flow of new pharmaceuticals. The idea is that start-ups will focus on research and pass promising compounds on to bigger companies capable of handling late-stage clinical trials and marketing, the cost of which can run into hundreds of millions of dollars.

If a biotechnology start-up succeeds, the major drug companies also sometimes like to buy the entire firm rather than just its products (*The Economist*, 2004a). For instance, MPM Capital, a Boston-based VC firm, which raised US$900 million in 2002, homes in on potential treatments for diseases of the central nervous system. It helped one of its portfolio companies, Hypnion, to raise US$47 million in order to support its search for gene(s) that control sleep so as more effectively to target a drug treating insomnia, particularly among America's rich and ageing population (*The Economist*, 2004a).

Like American venture capitalists, European firms are returning to better days, although they never suffered as big a bust. Europe is still a series of largely national markets for venture capital, and European start-ups tend to conquer these rather than trying to compete in global markets. Universities are becoming more reluctant to let academic research, whether in biology or engineering, mingle freely with commerce.

Disputes over the fruits of intellectual property are becoming more common, which is proving a barrier to VC investment. European venture capitalists also seem more risk averse than their American competitors. That is partly because banks and corporations still dominate early-stage investing. European VC funds are still hamstrung by tax rules and regulations; for instance, many pension funds are not allowed to invest in assets that are deemed too risky – venture capital included (*The Economist*, 2004a). European VC funds are more likely to be invested in building or refining technologies, rather than trying to develop an international brand or a company that has its own global sales force and marketing operations. Venture capitalists may increasingly fund start-ups using a portfolio approach in which individual VC firms team up with peers in order to pool risks (*The Economist*, 2004a).

In conclusion, experienced venture capitalists recall that during the boom they received more new business proposals than they could read, let alone evaluate sensibly. Although biotechnology investing is becoming fashionable again, discipline should be maintained and venture capitalists should not drop their guard because of an eventual flood of money (*The Economist*, 2004a).

Incentive and supportive measures in European countries

In 2004, the first steps were taken to improve the legal and fiscal framework for high-technology small and medium-sized enterprises (SMEs), such as most of the developing European biotechnology companies, and their access to capital. In January 2004, Belgium and France established special tax incentives for R&D-intensive companies. These were to boost start-ups, the growth and competitiveness of high-technology SMEs and investments in R&D. Meanwhile, the European Commission presented an "action plan", intended to support entrepreneurship and start-ups in Europe (Gabrielczyk, 2004). In Belgium, biotechnology and high-technology SMEs were to benefit from a 50 per cent reduction in income tax for company researchers who collaborate with public research institutions: the more the company invests in R&D, the more it will benefit from the tax exemptions to come into effect in 2005 (Gabrielczyk, 2004). The German government planned to establish a €500 million fund by 2005–2006 for high-technology SMEs. This, together with the creation of special seed funds, was aimed at bridging the gap in equity financing for most of the research-oriented biotechnology companies with products under development (Gabrielczyk, 2004).

Although a small group of pioneering biotechnology companies emerged in France in the 1980s, none of these first-generation companies is yet being considered successful. A large number of second-generation companies appeared, promising potential and robust dynamism. Indeed, the French bio-industry is rather heterogeneous and diverse, with a handful of mature companies and a burst of start-ups. Most of the current 260 biotechnology companies were founded between 1998 and 2002. A considerable number of the companies – most of which are predominantly (50 per cent) active in drug development – cluster around Paris (about 30 per cent). Other strong bio-incubators are located in Lyons, Strasbourg and Grenoble. There are developing bio-centres in Lille, Marseilles, Nantes, Bordeaux, Toulouse and Montpellier (Francisco, 2004).

The largest companies, which include companies listed on the stock market, employ an average of 100 people, have raised more than €50 million since their creation, have several products in development or on the market, and have sales of €5–100 million (e.g. Transgène, Nicox, Flamel Technologies, Cerep, which are all public companies; and Synt: em, IDM, Edonhit Therapeutics, Ethypharm Group, Biomedica Diagnostics, Proskelia Pharmaceuticals, Meristem Therapeutics, Genfit, Urogène and Proteus). Most of the companies do not earn revenues and one-third have a negative revenue growth rate, although 4 per cent of the companies account for 80 per cent of the sector's turnover. Sales and R&D spending are not always correlated in the early days of a company, which explains the long product development cycles, especially for therapeutic products (Francisco, 2004). Despite the economic crisis in France, the number of employees increased by 32 per cent in 2002, to more than 4,500 employees; R&D employees grew even faster (70 per cent in 2002). The employment of researchers in biotechnology companies is growing steadily – 48 per cent in 2002 (Francisco, 2004).

Biopharmaceutical companies (drugs and medical diagnostics) are primarily devoted to cancer research (19 per cent of therapeutic products in development), infectious diseases and diseases of the immune system and the central nervous system. Because most of the French companies started business in the late 1990s, most drugs are in pre-clinical trials (101), phase-1 trials (21) or phase-2 trials (31). Only seven products are undergoing phase-3 trials. This may be one reason bio-manufacturing capacities were also at an early stage in 2004. A study of France Biotech, the French bio-industry association, shows that production capacities have to grow when the biotechnology-derived products get closer to the market (Francisco, 2004).

The future of the French bio-industry depends strongly on an improvement in stock markets in the near future. Nevertheless, levels of venture funding for the industry have grown steadily. A dozen venture firms now have significant expertise in biotechnology and have dedicated personnel in Paris. Both the number and the quality of the business plans that are proposed to the venture capitalists have improved remarkably in recent years. Indeed, biotechnology and the life sciences became the priority sectors for venture capital investments in 2002. At the same time, the decrease in venture capital investments of 16 per cent compared with 2001, with a total of about €230 million invested, was felt in the French biotechnology sector. The decrease was more drastic in 2003, with only €90 million invested in the French biotechnology companies, making the economic environment tremendously tough again for the companies (Francisco, 2004).

Biotechnology companies tend to seek alliances with large pharmaceutical firms or groups to validate their technology in the eyes of their current and future investors, to fund R&D and decrease the need for dilutive rounds of equity financing, and to have a commercial partner that can sell their products effectively in markets that are not reachable without a large sales force. Three-quarters of the companies have research partnerships, both with academic laboratories and with other companies, especially with pharmaceutical laboratories. These partnerships, which represent the second source of financing for the companies, are mostly established in Europe (Francisco, 2004).

Investments in biotechnology SMEs share two characteristics: a high risk level and long-term returns on investment. To tackle these challenges, the 1999 Law on Innovation and Research introduced a new framework in France that has been making it easier for academic researchers to go into industry, has improved cooperation between public and private research, has started lightening the tax burden on innovative companies, and has improved their legal framework (Francisco, 2004).

However, the gap between Europe (including France) and the United States is widening: the investment ratio (e.g. venture capital, initial public offerings (IPOs), secondary offerings) between Europe and the United States is very low. Until October 2003, the ratio between Europe and the United States in terms of the amount of investment in biotechnology companies was 6 per cent. This very low ratio has two explanations: first, IPOs and secondary offerings started picking up again in the United States in 2003, whereas nothing happened in Europe; secondly, for younger companies, venture capital investments are lower in Europe than in the United States – with a ratio of about 20 per cent. According to Philippe Pouletty, president of France Biotech and of the Strategic Council for Innovation, and chairperson of the Emerging Enterprise Board of EuropaBio (the European Association of Bioenterprises), there is therefore a need to boost investment in European biotechnology companies. Otherwise the widening gap is going to make things very difficult for Europe (Pouletty, 2004).

It was essential to have major tax incentives to turn France into one of the most attractive countries for investors, scientists and entrepreneurs.

A new Innovation Plan was enacted by the French Parliament during the second half of 2003, for application as of 1 January 2004. France also adopted a new status for Young Innovation Companies (YiCs). Before this, France ranked last in Europe on operating costs; now, with the YiC status in place, France has moved to the top, having the lowest operating costs compared with other countries (Pouletty, 2004).

The first series of measures consists of specific support for the projects of the YiCs, defined as SMEs less than eight years old and bearing R&D costs of over 15 per cent of their total expenses. They can take advantage of a social and local tax break and their investors benefit from tax exemptions on equity added value. For a given SME, on average, it may end up increasing its available cash by one-third. The second series of measures is a tax credit linked to R&D expenses, based on their level and rate of increase. Nationwide, it is expected that total credit could double, up to €1 billion per year. In other words, the French government delivered a clear message: no taxes – with total exemption from social costs and salaries (which are extremely high in France, and in Germany too); total exemption from capital gains taxes; total exemption from corporate income taxes as well as local taxes – and this for a period of eight years following the creation of a corporation, as long as this corporation puts 15 per cent of its overall expenditures into R&D (Pouletty, 2004).

Combined with a reform of the tax research credit systems and a new status for Business Angels, this reform will bring true relief to the biotechnology industry. In addition, the announcement on 6 March 2004 by the French government of the creation of an agency dedicated to biotechnologies, with a budget of €3 billion and functioning like the US National Science Foundation, is considered a move in the right direction by the companies and industrialists.

In fact, there is a lot to be optimistic about in relation to French and European biotechnology. Governments have become increasingly supportive of the biotechnology sector and have passed laws that will impact on biotechnology and entrepreneurial activity. Companies are able to raise significant amounts of private equity and it is not rare to see second rounds of financing over US$20 million (Francisco, 2004). Apart from this, the creation of a European stock market needs to be pushed forward to attract European biotechnology SMEs. This should be the next target so that tax incentives can have a positive effect (Gabrielczyk, 2004).

The expiry of patents for biotechnology-derived drugs and its economic impact

The following biopharmaceuticals have lost or will soon lose their patent protection: Genzyme Corporation's alglucerase (2001); Eli Lilly's human insulin, Humulin (2003), and somatotropin, Humatrope (2003); Biogen's beta-1a interferon, Avonex (2003); Genentech's somatotropin, Nutropin (2003); Novo Nordisk's human insulin, Novolin (2005); Genentech's somatrem, Protropin, or human growth hormone (2005); Genentech, Boehringer-Ingelheim, Mitsubishi, Kyowa Hakko Kogyo's alteplase, Activase (2005) (Adhikari, 2004). The expiry of leading biopharmaceuticals' patents may unleash a flood of biogeneric counterparts. By 2006 therefore, 11 biotechnology-derived drugs, worth more than US$13 billion in annual sales, could face generic competition in the United States.

However, there are technical hurdles, because many of these biotechnology-derived drugs, unlike chemically derived ones, are complex proteins that can be synthesized only by using living cells. The process is prone to contamination and is highly variable, making the task complex and costly (*The Economist*, 2003c).

The marketing of bio-similar drugs can be also tricky. Most generics companies sell to pharmacists, who in many countries are allowed to dispense them as substitutes for expensive branded drugs. By contrast, biotechnology-derived drugs are mainly administered by physicians, and therefore the marketing of their generic versions will have to be redirected to doctors. But generics drug-makers lack the good relationship that many patent-holders have with physicians. According to Lehman Brothers (an investment bank) it could cost up to US$25 million to bring a bio-similar drug to market, 10 times more than a conventional generic. This means that competition in bio-similar drugs is likely to be limited – and hence the prices of the drugs will stay relatively high; some estimates foresee biotechnology-derived products selling for roughly 60–80 per cent of the price of the patented original. For conventional drugs, generics can easily sell for as little as 20 per cent of the branded price in the first year (*The Economist*, 2003c).

Unlike the costly and time-consuming clinical trials that big drug companies have to carry out with new drugs to win the approval of regulators in the United States and Europe, generics firms have to show only that their copy versions are chemically identical and behave in the human body in the same way. Regulators are wary about applying the same approval process to bio-similar drugs, where even the slightest difference in production can lead to subtle changes in a protein; this, in turn, can make it behave quite differently in the body. Consequently, the cost of additional clinical trials, as well as the close monitoring of products in the market, could greatly reduce the profits for bio-similars (*The Economist*, 2003c).

Given these hurdles, only a few generics makers are likely to go into that business. For instance, Sandoz AG, based in Vienna, hoped to be the first on the European market in 2004 with its generic version of human growth hormone. Sandoz is part of Novartis, and can draw on its parent company's expertise and funding for assistance. Alternatively, firms better placed to introduce bio-similars into Europe and the United States are those already working outside these key markets, in India and China for instance, with years of experience in manufacturing, testing and selling generic products in places where patents have not been strictly enforced. Firms such as Dr Reddy's and Wockhardt, in India, and Pliva, a Croatian generics maker, may be able to overcome the difficulties of producing and selling bio-similars and could later on gain access to the lucrative "western" markets (*The Economist*, 2003c).

5

Promising areas and ventures

Medicines from transgenic animals

Biotechnology companies that use transgenic farm animals as bioreactors to produce life-saving medicines are worth mentioning from both the technological and commercial viewpoints. If the manufacture of bio-pharmaceuticals in animals could be made more efficient, it would ease the flow of new therapeutic compounds, which is being held back. More than 100 protein-based drugs are currently in advanced phases of clinical trials, and many more are in development in the laboratory. So far, Gen-zyme Transgenics Corporation Biotherapeutics has successfully engi-neered goats that produce 14 varieties of therapeutic protein in their milk. Creating a flock of transgenic goats costs about US$100 million; this is expensive, but represents only one-third of the cost of building a protein-production facility. Furthermore, when a drug-maker needs to double production, the solution is to breed more animals rather than spend US$300 million on a new factory (*The Economist*, 2003a). It gener-ally takes about 18 months to make a transgenic goat that produces a de-sired therapeutic protein in its milk. For a cow this period is about three years, but milk production is higher (20 litres a day compared with 2 litres a day for a goat). Regarding the production cost of the purified therapeutic protein, this could fall to US$1–2 a gram, compared with US$150 a gram when the protein is extracted from cultured mammalian-derived cells (*The Economist*, 2003a).

Chickens have some advantages over goats and cows. They lay eggs that are sterile, and the albumen (egg white) is an ideal storage medium for fragile compounds. Chickens mature faster and are cheaper to breed than goats and cows; a chicken flock can multiply 10-fold within a year. Despite these advantages, research on transgenic chickens is less advanced than that on cattle or goats. In July 2002, TranXenoGen announced that it had produced two antibodies (one human and one murine) in the albumen of transgenic chickens. The yields of these antibodies need to be increased and it will take the company another year or so to achieve this. TranXenoGen also aimed to produce transgenic chickens laying eggs containing insulin and human serumalbumin (*The Economist*, 2003a).

The few corporations that are breeding transgenic farm animals with a view to producing medicines hope to pocket huge profits from sales estimated at billions of dollars over the next decade. But they might be outpaced by those companies that have chosen to carry out the same process in crops such as maize, alfalfa and potato at a lower cost.

Plant-derived drugs using molecular biology and biotechnology

The use of plant materials and extracts for medical purposes has a long history. Plants with relevant medicinal properties are identified (botany) and the corresponding active compounds are purified and tested for both efficiency and safety (biochemistry and medicine). Then the best approach to production is determined, whether chemical synthesis or extraction from natural sources (intact plants, or tissue or organ cultures), or a combination of both. A classic example of a plant-derived drug is salicylic acid, found in willow bark, which is the basis for acetylsalicylic acid, or aspirin, for the relief of pain, fever and inflammation, and more recently for protection against strokes. Aspirin – a simple molecule with no chirality – is produced via chemical synthesis on an industrial scale (Covello, 2003).

Another example, more complicated than aspirin, is taxol – an isoprenoid compound found in the bark of the Pacific yew tree (*Taxus brevifolia*). Taxol is very effective in ovarian and breast cancer treatment. It is a complicated molecule, whose total synthesis was reported in 1994 and involved over 40 steps. However, this was not commercially viable. On the other hand, yew bark contains only 0.01 per cent taxol, so supply is problematic. Production from cell cultures has been disappointing, although proprietary plant-cell fermentation technology recently developed may help. The best solution so far has been semi-synthesis. A precursor is obtained from yew needles and modified chemically to give taxol; this is not an entirely satisfactory solution and taxol is a prominent example of a plant-derived drug that is difficult to supply (Covello, 2003).

Another approach is the application of molecular biology and biotechnology tools. With better knowledge of the genes involved in the biosynthesis of plant-derived drugs, their production may be enhanced in plants, plant-cell cultures or alternative hosts. This metabolic engineering approach is an area of active research. For instance, Rodney Croteau and colleagues at Washington State University are tackling the taxol problem in this way. Part of the Canadian National Research Council's programme on crop genomics, led by Wilf Keller at the Plant Biotechnology Institute (PBI), involves molecular genetic approaches to two other classes of plant-derived drugs with similar supply problems – aryl tetralin lignans and tropane alkaloids.

Podophyllotoxin belongs to the class of aryl tetralin lignans. It is an anti-viral drug and it is chemically modified to give three related anti-cancer drugs – Etoposide, Etopophos and Teniposide. Etoposide, for instance, has extensive application in the treatment of small cell lung cancer, advanced testicular cancer and Karposi's sarcoma. Other podophyllotoxin-related compounds have shown promise as anti-HIV agents. However, these compounds have four contiguous chiral centres and a high degree of oxygenation, so total chemical synthesis is not commercially viable. Moreover, the current source of podophyllotoxin is a relatively rare herbaceous plant called *Podophyllum hexandrum* that grows in the Himalayas and whose current rate of harvest from the wild exceeds its regeneration rate (Covello, 2003).

A search for the relevant genes from *P. hexandrum* is being carried out at PBI. There is already fair knowledge about intermediates and enzyme-coding genes for the early part of the biosynthetic pathway from common phenylpropanoid precursors. Less is known about the later part of the pathway. The Canadian researchers start by making a library that represents the genes in the tissue that make the compound of interest. A few thousand are selected at random and sequenced. The sequences, called expressed sequence tags, are compared against a large sequence database to give tentative identifications. Using additional data, such as knowledge of gene expression in different tissues, the researchers can identify candidate cDNA clones that may encode enzymes of interest. These candidates are then tested by heterologous expression – the genes are introduced into a host such as *E. coli* or yeast, to test for activity from the enzyme of interest. The enzyme assays involved require good biochemical and analytical expertise because different and often unusual substrates and products are required for each assay. The enzyme assays may point to the cDNA candidates corresponding to genes involved in podophyllotoxin biosynthesis (Covello, 2003).

Once the genes have been found, it should be possible to engineer an appropriate host organism metabolically; a plant host may be the most feasible, but micro-organisms that can be easily contained may be more attractive than plants. Although one might wonder about the difficulty of introducing a number of genes into a plant simultaneously, this is becoming more common. For instance, 3 genes were introduced into rice for the production of beta-carotene ("golden" rice) and up to 12 have been successfully introduced into soybeans. If not all of the genes of interest in a biosynthetic pathway are found, individual genes may be put to use in various ways. One is to remove metabolic bottlenecks in plants or cultures; another possibility is the metabolic engineering of intermediate compounds with potential drug precursors (Covello, 2003).

Plants produce a wide range of secondary metabolites for defence and survival in their ecosystems. These secondary metabolites, currently exceeding 100,000 identified substances, belong to three major chemical classes: terpenes (a group of lipids), phenolics (derived from carbohydrates) and alkaloids (derived from amino-acids). The terpenes include the diterpene taxol from the Pacific yew and the triterpene digitalin from foxglove used as an effective drug for congestive heart failure. The phytoalexin resveratrol, an anti-oxidant agent, is an example of a phenolic, as are flavonoids and tannins, which are found in tea, fruits and red wine and have many desirable health effects. Alkaloids, a major class of plant-derived secondary metabolites used medicinally, have potent pharmacological effects in animals owing to their ability to penetrate cell membranes rapidly. Nicotine, a commercially important alkaloid, is the most physiologically addictive drug used by humans. Caffeine, an alkaloid from coffee, tea and chocolate, is a central nervous system stimulant and mild diuretic. Vincristine and vinblastine, alkaloids from periwinkle, are strong antineoplastics used to treat Hodgkin's disease and other lymphomas. The opium plant contains over 25 alkaloids, with morphine being the most abundant and most potent painkiller (Oomah, 2003).

Because of the numerous applications of plant extracts and isolated secondary metabolites, their world market exceeds US$10 billion annually. The pharmacological value of plant secondary metabolites is increasing owing to constant discoveries of their potential roles in health care and as lead chemicals for new drug development. However, secondary metabolites, generally present at 1–3 per cent of dry plant biomass, are synthesized in specialized cells at distinct development stages and have highly complex structures, making their extraction and purification difficult. The content of secondary metabolites in biomass has been increased by cell-culture techniques in an effort to facilitate large-scale production, although the economic efficiency remains questionable (Oomah, 2003). In future, the idea is to apply molecular biology and biotechnology tools to better understand the biochemical pathways leading to these compounds, isolate the genes coding for the relevant enzymes, and metabolically engineer the producing plants or alternative hosts in order to increase the synthesis and output of secondary metabolites (Oomah, 2003).

The opium poppy (*Papaver somniferum*) has the unique ability to synthesize morphine, codeine and a variety of other benzylisoquinoline alkaloids (including papaverine and sanguinarine) that are of pharmaceutical importance. The global market for licit opium, from which the alkaloids are extracted, was in excess of 160 tons annually in 2003. The opium poppy produces small amounts of codeine because demethylase activity converts codeine into morphine. Approximately 95 per cent of the morphine extracted from licit opium is chemically converted into codeine, which is a more versatile pharmaceutical. The large quantities of morphine produced by the plant are the basis for the illicit cultivation of the opium poppy in many regions of the world in order to synthesize 0,0-diacetylmorphine, or heroin. The illicit production of opium is almost 10 times greater than licit production (Facchini, 2003).

Research aimed at metabolic engineering in the opium poppy could lead to biological alternatives that would reduce the production and trafficking of illicit drugs around the world. Moreover, improved knowledge about secondary metabolism in the opium poppy could create opportunities to introduce entire pathways into value-added crops. In the case of benzylisoquinoline alkaloid biosynthesis in the opium poppy, considerable progress was made during the 1990s. No fewer than eight genes encoding alkaloid biosynthetic enzymes have been cloned from the opium poppy. However, the biosynthesis of morphine and related alkaloids, such as the antimicrobial agent sanguinarine, involves many more enzymes. In addition, much remains to be learned about the control of alkaloid biosynthetic pathways, which are strictly regulated in plants. Consequently, our ability to harness the vast potential of these important secondary pathways is still rather limited. For instance, the use of plant organ, tissue and cell cultures for the commercial production of pharmaceutical alkaloids has not become a commercial reality despite decades of empirical research. Plant-cell cultures of the opium poppy can be induced to accumulate sanguinarine, but they do not produce morphine. This inability of dedifferentiated cells to accumulate certain metabolites has been interpreted as evidence that the operation of many alkaloid pathways is tightly coupled to the development of specific tissues. Recent studies have shown that alkaloid pathways, in general, are regulated at multiple levels, including cell type-specific gene expression, induction by light elicitors, enzymatic controls and other poorly understood metabolic mechanisms (Facchini, 2003).

Advances in genomics will provide a more rapid and efficient means to

identify the biosynthetic and regulatory genes involved in alkaloid pathways. The importance of a multifaceted approach to studying the regulation of alkaloid biosynthesis in plants such as the opium poppy is highlighted by novel insights obtained using a combination of conventional
and modern techniques, including biochemistry, molecular biology, cell
biology and genetic engineering, as well as advanced chromatographic
methods to isolate, characterize and assess secondary metabolites rapidly
and accurately (Facchini, 2003).

Biopharming

Plant-based pharmaceuticals

Epicyte Pharmaceutical is one of a host of biotechnology companies involved in the production of plant-based pharmaceuticals. Researchers
have launched more than 300 trials (2003) of genetically engineered
crops to produce everything from fruit-based anti-hepatitis vaccines to
drugs against HIV/AIDS in tobacco leaves. Open-air trials of pharmaceutical crops (the process is called biopharming, which can be linked to
medical biotechnology or to agricultural – "green" – biotechnology) have
taken place in 14 US states, from Hawaii to Maryland. Clinical trials have
begun for experimental crop-grown drugs to treat cystic fibrosis, non-
Hodgkin's lymphoma and hepatitis B (*The Economist*, 2003a). Many researchers in North America, Europe and a few other parts of the world
are working to develop plant-based production systems for human therapeutic proteins.
   Crohn's disease is a chronic inflammatory bowel disease with dis-
continuous lesions occurring throughout the intestinal tract. Current
treatments, which rely heavily on corticosteroids and immuno-solocilates,
have multiple side-effects and complications. There is therefore a need
for alternative therapies. Indications that cytokines retain some of
their biological activity following oral administration, coupled with
improvements in patients following parental administration of the anti-
inflammatory cytokine interleukin-10 (IL-10), led researchers at the
Southern Crop Protection and Food Research Center to believe that
oral IL-10 may be an effective means of delivery to the gut. In order to
produce sufficient low-cost IL-10 for oral administration, these researchers have expressed the human gene for IL-10 in low-nicotine tobacco.
They have evaluated the biological activity of transgenic plants and intend
to use them in an animal model of Crohn's disease (Menassa et al., 1999).
   William Langridge (professor in the department of biochemistry and at
the Center for Molecular Biology and Gene Therapy at the Loma Linda

University School of Medicine, California) and his colleagues have used transgenic potatoes to synthesize human insulin and showed that diabetic mice fed with these potatoes were less affected by the disease and their symptoms regressed. Douglas Russell of Monsanto obtained tobacco plants producing human growth hormone (somatotropin) in its biologically active form. And tomatoes producing an inhibitor of the enzyme converting angiotensin-1 (i.e. an anti-hypertensive compound) were produced. Other plant-produced substances include: enkephalins, alpha-interferon, serumalbumin and two of the most expensive pharmaceuticals – glucocerebrosidase and the factors stimulating the colonies of granulocytes and macrophages.

Glucocerebrosidase is an enzyme whose function is markedly decreased in patients suffering from Gaucher disease, which causes mental retardation in children, the inhibition of large bone growth and an increase in size of the liver and spleen. Another protein produced commercially in transgenic plants is hirudin, an anti-coagulant protein extracted from leeches, used to treat thrombosis. It has been produced in colza and mustard seeds by SemBioSys Genetics, Inc. Another commercial production example is human alpha 1-anti-trypsin in rice seeds by Applied Phytologics. This protein is used in the treatment of cystic fibrosis and haemorrhagia. Applied Phytologics hoped to obtain authorization for commercializing this product in 2004.

Lactoferrin is the second most abundant protein in human milk, present at a concentration of 1 gram per litre. It is a multifunctional protein that protects against microbial infection caused by a broad spectrum of bacteria and may have an important role in the regulation of iron uptake in the gastro-intestinal tract as well as in the regulation of systemic immune responses through cytokine release. Fogher et al. of the Botany and Genetics Institute of the Catholic University S. Cuore, Piacenza, and Plantechno Srl, Cremona, Italy, have designed and transferred a synthetic lactoferrin gene into rice, with the aim of engineering a functional food to be used either directly or as a vegetable milk in human nutrition. The synthetic human gene was optimized for codon usage in plants: the researchers used the regulatory elements of the soybean storage proteins 7S globulin and beta-conglycinin, with their own specific leader sequences, as seed-specific promoters. The transgenic rice lines were controlled for seed-specific lactoferrin production, iron content of the seeds and glycosylation pattern of the protein. Using the two rice varieties Ariete and Rosa Marchetti, an 82 kilodalton protein recognized by the anti-human lactoferrin antibody was produced in the seeds and not in the leaves. The amount of the protein accumulated exceeded 1 per cent of total seed proteins, and the plant lactoferrin was glycosylated at the same level as in the human protein. In some transgenic rice lines, the quantity of seed iron was three times the normal concentration in the non-transgenic variety (Fogher et al., 1999).

SemBioSys Genetics is using a variety of genetic engineering techniques to express proteins in the seeds of safflower. One embodiment of this technology involves the covalent attachment of proteins to oil bodies, natural oil-storage organelles found in oilseeds. Taking advantage of the physical principle that oil is lighter than water, oil bodies can be easily separated from the majority of other seed components. This provides a cost-effective solution for bulk protein production and purification. The technology is amenable for oral and topical delivery of bioactive peptides and proteins. The company operates a manufacturing facility that can deliver oil-body-based products and purified protein at scale. SemBioSys Genetics has been granted a US patent covering the use of its technology to produce somatotropins including human, bovine or fish forms of the hormone. This was the first example of the commercial-level expression of this class of proteins (correctly folded and active disulfide-bonded proteins) in seeds. The fish somatotropin, thus produced, was physiologically active as an oil-body-associated protein when fed to salmon and trout, demonstrating its potential for oral delivery of biologically active proteins.

Plantibodies and vaccines

Although antibodies were first expressed in plants in the mid-1980s, the first report was published in 1989. Since then, a diverse group of "plantibody" types and forms have been prepared. Originally, foreign antibody genes were introduced into plant cells by non-pathogenic strains of the natural plant pathogen *Agrobacterium tumefaciens*, and regeneration in tissue culture resulted in the recovery of stable transgenic plants. Although this initial work to generate multi-chain proteins required the crossing of plants expressing each chain, further studies have shown that multiple chains can be introduced via a single biolistic transformation event, greatly reducing the time to final assembled plantibody. James W. Larrick and his colleagues at the Palo Alto Institute of Molecular Medicine and of Planet Biotechnology have transformed tobacco plants so that they produce antibodies against one of the surface proteins of *Streptococcus mutans* (the agent of tooth decay); these plantibodies could successfully prevent the recolonization of the teeth and gums by this major oral pathogen in patients treated with the antibodies for at least four months (Larrick et al., 2000).

At Thomas Jefferson University, the work carried out by Hillary Koprowski has shown that persons fed with transgenic lettuce containing an antigen of the hepatitis B virus had a good immune response against that virus. In 1998, Kevin Whaley and his colleagues, using soybean plants transformed to produce antibodies against herpes virus simplex type 2, succeeded in preventing vaginal infection by this virus in mice that had eaten the plants. These plantibodies were 100 to 1,000 times more efficient than other products used previously in this kind of experiment.

William Langridge is leading a team that is genetically engineering potatoes to provide vaccination against cholera. The disease is contracted by more than 5 million people annually and causes more than 200,000 deaths worldwide. The researchers succeeded in transforming potatoes to produce the beta subunit of cholera toxin (CTB). These potatoes were then used in two experiments. In the first, the team fed uncooked transformed potatoes to mice to develop the anti-CTB immunity. These mice were found to contain cholera antibodies in their blood and faeces. In addition, those fed the biggest amount of potato were shown to be more resistant to the cholera toxin. When the effects of the vaccine wore off, the mice were given a booster to maintain their immunity. In the second experiment, the researchers boiled the potatoes. An analysis of the boiled tissues showed that 50 per cent of the CTB remained (Langridge, 2000).

It is already known that CTB gives greater immunity against cholera to humans than it does to mice. However, existing vaccines do not adequately protect against this disease. The difficulty lies in the method of immunization. Vaccines are generally injected and stimulate an immune response in the bloodstream. But injections do not produce antibodies on mucosal surfaces, e.g. the walls of the gut. Vaccinations with plant-derived vaccines (instead of calling them "edible" vaccines) offer an alternative strategy to tackling this type of infection. Oral administration of transgenic-derived products will deliver cholera vaccine directly to the gut and provoke an immune response precisely where it is wanted. Langridge and his colleagues have yet to take their technology out of the laboratory (Langridge, 2000).

Researchers at the Baylor College of Medicine in Houston have carried out trials on transgenic potatoes designed to protect against Norwalk virus, a major cause of water- and food-borne diarrhoea in developing countries. Vaccine-containing potatoes developed at the Boyce Thompson Institute for Plant Research, an affiliate of Cornell University, by Charles Arntzen and Hugh Mason were also used in a human clinical trial. In this trial, led by Carol Tackett at the University of Maryland School of Medicine, volunteers were each served three helpings of raw potato to test the use of this kind of vaccine against traveller's diarrhoea. This common condition is transmitted by food or water contaminated by enterotoxigenic *Escherichia coli*. The results of the trial, presented in the journal *Nature Medicine*, showed that individuals who had eaten the transgenic potato developed antibody protection against the diarrhoea.

"Since infectious diseases cause a loss of more than 15 million kids annually, much of which could be prevented by good vaccines, it is our obligation to explore all new approaches to inexpensive and effective disease prevention", stated Charles Arntzen.

Charles Arntzen has also been working for nearly five years to create transgenic tomatoes containing a gene from a strain of *Escherichia coli* that can protect against diarrhoeal diseases. The US researcher focused on diarrhoea, because these diseases kill at least 2 million people in the world annually, most of them children. And he chose tomatoes because greenhouse-grown tomatoes cannot easily pass their altered genes to other crops and because tomato-processing equipment is relatively cheap. It would be easier to eat whole tomatoes, but that would be a disaster, according to Charles Arntzen. Individual tomatoes come in different sizes with varying concentrations of the new protein, whereas uniformity of dosage is the key to an effective vaccine. In 2002, Arntzen tested juice derived from transgenic tomatoes on animals, with human trials to follow (Lemonick, 2003).

In relation to the research carried out at the Boyce Thompson Institute for Plant Research on plant-derived vaccines, studies have been made on the elements in banana that promote storage proteins, with a view to manipulating them for better delivery of the vaccine. Clendennen et al. (1998) studied a banana protein (P31) that appears to have evolved from an enzyme. P31 is abundant in the pulp of green bananas and it may play a role as a storage protein. P31 belongs to a class of proteins called the class III chitinases, several of which, although they are present during fruit ripening in avocado, cherry and tomato, have been thought to protect plants from disease or wounding; in certain plants, they are known to protect against fungi. In the banana, however, P31 chitinase decreases in abundance as ripening progresses. In general,

- storage proteins are very abundant (in unripe banana pulp, P31 accounts for approximately 20–30 per cent of total soluble pulp protein);
- storage proteins are broken down during a subsequent developmental stage (P31 is broken down during banana ripening);
- storage proteins are generally localized in storage vacuoles within the cell (P31 is localized there);
- storage proteins contain a great proportion of particular amino-acid residues (P31 contains 22 per cent of such residues, approximately the same as in soybeans and poplar-storage proteins – 21–25 per cent);
- storage proteins typically lack any other metabolic or structural role in the organism; some retain a little enzymatic activity, but this could not be the case for the banana protein, which has only three of the five amino-acids required.

Clendennen et al. (1998) suspected that P31 serves as a storage protein in banana and carried out experiments to determine whether a P31-promoting element introduced into tomato plants by genetic engineering would prove to be related to fruit ripening in the tomato, supporting the view of the role of P31. Storage-protein-promoting elements might be genetically engineered to assist in vaccine delivery.

Some four dozen laboratories around the world are working on their own versions of plant-derived vaccines, using tomatoes, bananas and potatoes. The advantages of such vaccines, particularly in developing countries, are their cheap cost and their oral consumption instead of using needles (thus avoiding contamination resulting from the lack of rigorous asepsis). Attention is focused on tomatoes and bananas, which look likely to be more suitable for subtropical and tropical regions. Axis Genetics is working in partnership with the Boyce Thompson Institute for Plant Research. The aim is to market vaccines in preserved foodstuffs or tablets against traveller's diarrhoea, Norwalk virus and hepatitis B.

With regard to the regulatory aspects of plant-derived vaccines, an important consideration is that these vaccines must never occur as normal constituents of food, and should be produced under regulatory conditions to prevent contamination of food supplies and to maintain genetic containment. Furthermore, the vaccines will not be delivered in fresh form, but will be processed to yield uniform, stable batches with well-defined antigen content (hence the preference for food tablets containing the appropriate doses of antigen, rather than the raw genetically modified vegetable or fruit). The vaccines will also be delivered by health-care professionals. Use of plants that are infertile and clonally propagated could facilitate management of quality control and production. The creation of male-sterile lines will enable more rigorous containment, because maternal inheritance of the chloroplast genome prevents pollen-mediated gene flow (Mason et al., 2002; Walmsley and Arntzen, 2003).

Charles Arntzen, now at Arizona State University, foresees rich markets for plant-derived vaccines to protect fish and poultry against diseases currently being treated – and in many cases over-treated – with conventional antibiotics. For instance, clinical trials carried out by the biotechnology company ProdiGene have shown that the feeding of pigs with transformed maize containing a vaccine against transmissible gastroenteritis could protect them effectively against this disease (Roosevelt, 2003).

Rinderpest is an extremely contagious disease of cattle, buffaloes, sheep, goats and wild ruminants, with a high mortality rate. The rinderpest virus has only one antigenic type (serotype) and an attenuated, live vaccine with high immunogenicity is available. Rinderpest has been eradicated in developed countries, but is still prevalent in parts of Africa, the Middle East and South Asia, where eradication campaigns are under way. The major drawback of the currently used vaccine against rinder-

pest is its heat lability. In hot countries, the vaccine delivery is con-
strained by high costs and the lack of maintenance of refrigeration
throughout the whole chain of production and distribution to keep the
potency of the vaccine. Although recombinant vaccines – vaccinia/
capripox recombinant or bacculo recombinants – have been produced,
they have not been tested in the field nor has their usefulness in providing long-term immunity been experimentally proven. Khandelwal et al.
of the Department of Microbiology and Cell Biology of the Indian Institute of Science have undertaken to develop transgenic tobacco plants
(model system) and transgenic groundnut (*Arachis hypogea*) plants expressing the haemagglutinin (H) protein of rinderpest virus, as a source
of vaccines to be delivered through feedstuffs in order to immunize domestic ruminants as well as susceptible wildlife. The expression of H protein was demonstrated in the transgenic plants, and about 250 transgenic
groundnut lines have been obtained by the Indian researchers (Khandelwal et al., 1999).

Preferred crop species

So far, more than two-thirds of plant-based medicines are being tested in
maize – a crop whose genetics is well understood. At Epicyte Pharmaceutical's laboratory, tiny tobacco leaves, transformed with herpes antibody
genes, were grown in incubators.

The Sacramento-based biotechnology company Ventria Bioscience
broke new ground by planting 130 acres with new varieties of transgenic
rice that will produce lactoferrin and lysozyme, to be marketed for use in
oral rehydration products to treat severe diarrhoea. The company stated
that this acreage could generate sufficient lactoferrin to treat at least
650,000 sick children and enough lysozyme for 6.5 million patients. It
hoped to expand production to 1,000 acres within a few years (Lean,
2004).

The company did not disclose the site earmarked for the new crops because it was worried that protesters might destroy them. However, its
plans have caused alarm in California's rice-growing community. Organic
farmers, in particular, feared that transgenic rice could contaminate their
crops. On 29 January 2004, the arguments were thrashed out before a
meeting of the California Rice Commission, which was drawing up a protocol of conditions under which the transgenic rice varieties could be
grown. In particular, the Commission was focusing on working out precautionary measures, e.g. the distance transgenic rice must be from conventional crops, to try to minimize the risks (Lean, 2004).

In the case of water lentils, LemnaGene LLC is specializing in genetically transforming plants of the Lemnaceae family through an agreement
concluded with Bayer CropScience. Based in Oregon, LemnaGene is col-

laborating with the Weizmann Institute of Science and the Yeda Research and Development company in Israel. Transgenic plants will be used to manufacture functional foodstuffs and new molecules for industrial, pharmaceutical and cosmetic uses. The advantages of *Lemna* spp. are their high productivity, good knowledge of their genetics and transformation process, and the possibility of growing them hydroponically in greenhouses (rather than in the open air), to avoid the escape of transgenes into the natural or agro-ecosystems. Because it is not a food or industrial crop, the transformed water lentil cannot "contaminate" conventional crops and may be preferred to maize for the production of drugs or other materials.

Comparative economic advantages

Biopharming is mostly driven by a cost advantage. Building sophisticated factories to produce biopharmaceuticals can take as long as seven years and cost up to US$600 million per facility. It is predicted that medicinal products could be synthesized in plants at less than one-tenth of the cost of conventionally manufactured drugs and vaccines. These costs are, for instance, 10–50 times less than for protein produced at high concentration in *Escherichia coli* (i.e. 20 per cent of total protein). Depending upon the use of the protein and the requirements for purification for *in vivo* pharmaceutical use, purification costs will obviously add to the final product costs; however, at the 100–1,000 kg level, plant-produced proteins will provide obvious savings. Consequently, demand for these low-cost products would grow rapidly and would far outstrip the capacity of conventional systems of production. For example, whereas pharmaceutical factories produce interleukin-10 in kilogram quantities, farmers would be able to produce it literally by the ton.

By the end of the current decade, biopharmaceuticals are projected to grow into a US$20 billion industry. How many of the new drugs will be manufactured in plants remains uncertain, however. This technology could bring down the cost of treating a number of diseases in a significant way, so that the drawbacks will be very small compared with the benefits (Roosevelt, 2003).

Pharmaceuticals and nutraceuticals from marine organisms

The large pharmaceutical groups seem to have abandoned their search for new drugs derived from natural substances: over the 1990s only one out of 10,000–20,000 compounds extracted from terrestrial microorganisms, plants or animals became an effective medicine. This may not be the case for marine organisms, as stated frequently by José María Fernández Sousa-Faro, who founded the first Spanish pharmaceutical company PharmaMar (and the only one up to 2004), a subsidiary of the Zeltia group, which is devoted to seeking new drugs from marine organisms. It is among the half-dozen companies across the world that carry out this kind of research, and to that end it has close ties with the fisheries group Pescanova, which owns fishing vessels and screening centres in every ocean (Pujol Gebelli, 2003). Located near Madrid, PharmaMar has been working for 17 years on the research and development of anti-tumour products derived from marine organisms. During the three-year period 2000–2002, its staff increased from 70 to 300. Its new facilities, opened in March 2003, required a €22 million investment and included a pilot plant for the production of drugs.

In 1996, PharmaMar's president announced that a compound derived from a marine Tunician, ET-743 (ecteinascidin-743), was to be tested in clinical trials for its potent anti-tumour activity. Two years later, at the Congress of the European Organization for Research and Treatment of Cancer (EORTC) in Amsterdam, Fernández Sousa-Faro reported the initiation of phase-2 clinical trials of this compound, renamed Yondelis. It became the the company's star product (Pujol Gebelli, 2003).

At the same time, PharmaMar went public and was able to raise €230 million on the stock market, of which €120 million was devoted to developing the first Spanish anti-tumour compound. This was very unusual in Spain, where companies never fund their research and development (R&D) with the money raised on the stock market. NeuroPharma and PharmaGen – PharmaMar's subsidiaries – are willing to follow suit, which is very good news for the Spanish R&D system, which lacks really innovative firms. In the United States, by contrast, it is common to use funds raised on the stock market to support innovative research. In Spain most corporations have to rely on the public assistance provided by the Centre for Technological and Industrial Development, the Interministerial Commission on Science and Technology, the Ministry of Science and Technology's Programme for the Promotion of Technological Research, as well as the European Regional Development Funds (Pujol Gebelli, 2003).

By 2004, some €300 million had been invested in PharmaMar, of which €120 million was allocated to research on Yondelis. During that period, 4 products were tested in clinical trials and another 14 were at the preclinical stage. The €230 million raised in 2000 on the stock market fuelled the research and, by the end of 2003, €130 million was left for funding it for another two years. PharmaMar's president was hoping that during this period a breakthrough would occur regarding Yondelis (Pujol Gebelli, 2003).

Unfortunately, the European Agency for the Evaluation of Medicinal Products (EMEA) gave a negative judgement in the first evaluation phase of Yondelis, despite the fact that PharmaMar presented all the details concerning the anti-tumour activity of Yondelis at the EORTC congress in Amsterdam, and despite the failure of the two drugs generally used against sarcomas of soft tissues, doxorubicine and phosphamid. Over the 20-year period up to 2003, almost 50 products had been tested against this kind of sarcoma, without success; Yondelis was the first compound that demonstrated a specific activity. Soft-tissue sarcoma is a rare disease, representing only 1 per cent of all forms of cancer, which implies that it is not easy to find patients on whom any potential drug could be tested (Pujol Gebelli, 2003).

At the start of the phase-2 clinical trials, three patients died. Because 50 per cent of the product is metabolized in the liver and the rest is eliminated through the bile, patients with obstructed bile ducts did not eliminate the product, so that the dose was too high to be tolerated. It is easy to find out whether or not a patient has a non-obstructed bile duct (through a test of alkaline phosphatase and bilirubine); if not, the dose can be reduced. With this change in the treatment protocol, there were no problems. In addition, there was no cumulative toxicity, and some patients have received more than 30 cycles, whereas with the other available drugs one cannot go beyond 6 cycles (Pujol Gebelli, 2003).

Despite the setback at the EMEA level, PharmaMar's president remains optimistic about the future of the company's star product. In addition to its activity against soft-tissue sarcoma, Yondelis was found also to have a specific activity against ovarian cancer, particularly among women whose cancer is resistant to the usual treatments with taxanes and cisplatinum. Yondelis could be an alternative therapy and new studies were being designed. Despite this blow, PharmaMar is being consolidated and is really innovative in its development of drugs derived from marine organisms. New compounds that also have a novel anti-tumour mechanism may pave the way for new modes of action against tumorous cells. The near future will tell if PharmaMar can succeed in what is generally considered a risky venture.

New Zealand has an extensive coastline and diverse and unique algal flora, including 800 known species of seaweeds. This represents an ample source for exploration. Industrial Research Limited (IRL) and New Zealand Pharmaceuticals (NZP) formed a strategic alliance with the Hobart-based company Marinova Pty Limited, with a view to commercializing compounds derived from marine organisms for pharmaceutical and nutraceutical applications. The IRL was set up in 1992, following the restructuring of the Directorate of Scientific and Industrial Research. This restructuring resulted in the formation of several Crown Research Institutes. The IRL is owned by the New Zealand government and overseen by a board of directors, which includes representatives from several high-profile companies based in New Zealand. The IRL undertakes contract R&D projects in technology areas, including advanced materials, biochemical engineering, and complex measurements and analysis, alongside commercialization activities. It also has the capability of specialty manufacturing within IRL Biopharm and IRL Glycosyn. The carbohydrate chemistry team at IRL is recognized as a world leader in the development of high-value carbohydrate compounds (Boyd, 2003).

With new facilities, backed by 20 years in seaweed research, the development of novel compounds has now become commercially viable through the strategic alliance between New Zealand and Tasmania. Thus, NZP now manufactures Marinova's bioactive compounds and is a shareholder in Marinova. This close relationship brings significant expertise in process development under good manufacturing practice standards. The focus of the seaweed programme of Marinova, IRL and NZP is on high-value products. Initial commercialized products are for the dietary supplements market, with the prospects of launching botanical drugs in the short to medium term under a new US Federal Drug Administration (FDA) category (Boyd, 2003).

The research focus to date has been the polysaccharides (derived from hydrocolloids) in seaweeds (often sulphated galactans), which have a long history of being used commercially. Agars, carrageenans and alginates are all useful compounds that cannot be made artificially, so they have to be extracted from the cell walls of seaweeds. The New Zealand government has recognized the potential for creating a new high-value industry from seaweeds. This is reflected in a commitment of NZ$7 million over six years to the IRL, leveraging the substantial investments of Marinova, NZP, other research providers and industry stakeholders (Boyd, 2003).

Late in 2002, following five years of development, Marinova, IRL and NZP commercialized a novel compound – a polysaccharide from the seaweed *Undaria pinnatifida*, an introduced species harvested in Tasmania. Native to Japan and parts of Asia, where it is commonly known as wakame, *Undaria* has been introduced to New Zealand, South America and parts of Europe. Marinova, through its subsidiary Marine Resources Pty, developed a commercial harvesting and processing model built around containment of this invasive species of seaweed. The Tasmanian Department of Primary Industries, Water and Environment, together with the Department of Economic Development, has been fostering this new industry. Commercial harvesting of *Undaria* began in late 2002 and over 200 tons of biomass were harvested by divers during the first commercial harvest season, compared with previous pilot quantities of up to

60 tons. "Viracle with GFS™" was Marinova's first product, entering the US market as an anti-herpes therapy. Galacto Fucan Sulphate (GFS™) was proven a potent anti-viral *in vitro* and in human pilot trials. Marinova has made considerable investment in clinical research, trials and regulatory compliance with a number of markets, including Canada, the United Kingdom, Australia and New Zealand. The company planned to have a presence in these markets in 2004 (Boyd, 2003).

The focus of the Tasmanian and New Zealand researchers includes an evaluation of marine biological diversity, marine pest strategies, marine farming of high-value seaweeds and biosafety. NZP has been purifying biochemicals from natural raw materials since the early 1970s. It supplies a range of biochemicals to the pharmaceutical, cosmetic, health food and biotechnology industries and over of 95 per cent of its products are exported. NZP has changed its focus from the production of biochemicals from meat industry by-products, to those derived from plant materials. NZP has a long history of manufacturing polysaccharides, including heparin and the related glycosaminoglycan, chondroitin sulphate. Experience in the extraction and purification of these animal-derived polysaccharides led the company to develop new technologies for extracting marine polysaccharides in conjunction with Marinova. NZP was able to quickly convert the initial information provided by Marinova and IRL into a commercial extraction process for *Undaria* harvested in Tasmania (Boyd, 2003).

New Zealand's commitment to creating a high-value industry from seaweeds highlights the importance of biotechnology to this country, as shown in the report "Growing the biotechnology sector in New Zealand – A framework for action", presented to the government and released to the public on 6 May 2003 by the Biotechnology Task Force. The Task Force was set up under the New Zealand government's Growth and Innovation Framework in 2002 and comprised members from business, universities and Crown Research Institutes. The Task Force's report, produced in cooperation with the biotechnology sector, Industry New Zealand and the Ministry of Research, Science and Technology, outlined actions for commercializing biotechnology innovations, building the critical mass of the sector and removing barriers to growth. Key targets in the report included:

- creating an industry with NZ$10 billion market capitalization;
- tripling the size of the New Zealand biotechnology community from 350 to over 1,000 organizations;
- increasing total cluster employment from around 3,900 to over 18,000;
- raising the number of core biotechnology companies from 40 to over 200.

The report also stressed a global focus by biotechnology business and action by government to increase international awareness of New Zealand's strengths. The website Biospherenz.com – a joint initiative between the New Zealand government and bio-industry – was designed to provide information aimed at international investors and researchers.

An independent study by the Channel Group identified nine key biotechnology sectors that offered great potential for growth in New Zealand: biopharmaceuticals, bio-manufacturing, agricultural biotechnology, transgenic animals, bioactives, industrial and environmental biotechnology, nutraceuticals, and clinical research and trials.

Cosmeceuticals

A new market niche known as cosmeceuticals – products that are marketed as cosmetics but contain biologically active ingredients – is proving to be lucrative. Cosmeceuticals are indeed one of the fastest-growing segments of skin-care business: in 2003, total skin-care sales in department stores grew 6 per cent, whereas sales of cosmeceuticals and clinical brands jumped 77 per cent; in 2002, total skin-care sales were up 4 per cent, whereas sales of cosmeceuticals and clinical brands rose 62 per cent. Cosmeceuticals are so appealing because they are more affordable than Botox or Restalyne, which are injected into facial muscles to erase wrinkles, and yet enjoy medical credentials. According to Aurelian Lis, co-founder of Prescribed Solutions (a cosmeceuticals company), "cosmeceuticals offer some of the benefits of pharmaceuticals but are still inherently cosmetics" (Foster, 2004).

Big cosmetic groups have enlisted the help of dermatologists. Lancôme, a division of L'Oréal (the world's biggest cosmetics group, with 18.7 per cent of the €70 billion global cosmetics market and about €14 billion sales in 2003), hired a specialist in dermatologic laser surgery as a consultant in September 2003, and Prescriptives, a unit of Estee Lauder, recruited a dermatologist in October 2002. In 2003, Estee Lauder bought the Rodan & Fields skin-care line developed by two dermatologists. Virginia Lee, US research analyst at Euromonitor International (a market research group), stated that cosmeceutical manufacturers are positioning their products as an option before more drastic steps such as plastic surgery, Botox injections or chemical peels (Foster, 2004).

As cosmeceuticals have become more mainstream, some products have appeared in stores in addition to being available in physicians' offices. NV Perricone and MD Cosmeceuticals are found in upscale retailers. Sephora, owned by the world's largest luxury goods group, Louis Vuitton-Moët-Hennessy, is at the forefront of the cosmeceutical trend; its US stores sell several cosmeceutical brands, including NV Perricone, Dr

Murad and Dr Brandt. Sales of NV Perricone more than doubled in 2002 from US$11.9 million the year before to US$42.4 million (Foster, 2004).

Overall, cosmeceutical sales are rising. A report by the Freedonia Group, a market research firm, stated that the US$3.4 billion cosmeceutical industry was poised to grow 8.5 per cent to US$5.1 billion by 2007. Although it is a fraction of the US$33 billion cosmetics and toiletries market in the United States, cosmeceutical sales are growing much faster than the overall market. This trend has gained momentum owing to ageing baby boomers who want to look young and also to FDA approval of Botox in 2002. More than 1.6 million people in the United States removed wrinkles with injectable treatments in 2002, according to the American Society of Plastic Surgeons, and more than 4 million Americans had non-surgical cosmetic treatment. Allergan, which makes Botox, reported that fourth-quarter sales of Botox rose more than 20 per cent to US$158 million in 2003. For 2004, the company expected Botox sales to grow to US$660–700 million – up as much as 24 per cent from US$563.9 million in 2003 (Foster, 2004).

6

Medical and pharmaceutical biotechnology in some developing countries

In a number of developing countries – those with advanced research and development (R&D) in the life sciences – public research institutions and biotechnology companies have invested in biotechnology R&D (medicine, agriculture and environment) and represent success stories at national and even regional level.

Argentina

The family-owned pharmaceutical company Sidus, based in Buenos Aires, made the decision to invest in medical biotechnology in the early 1980s. It created a subsidiary, BioSidus, which started producing its first biotechnology-derived product – recombinant erythropoietin (EPO) – at the end of the 1990s. The company recruited a first team of five academic scientists and set up a pilot plant to produce EPO, with other products to follow. In 1993, BioSidus achieved financial autonomy; in other words, Sidus had been investing money in its subsidiary for 13 years before it became profitable. The investment has been estimated at about US$35 million, which can be considered as venture capital responding to a long-term vision for development. Nowadays, BioSidus has a 10-year business plan and much more investment is needed, which will probably require the company to go public and be quoted on the stock market.

The company's organization has demonstrated that, as in the developed countries, it is not just good scientists who are needed, but also specialists in marketing, finance, law and regulation. This combination of competences is the key to successfully transforming research results into a marketable product. BioSidus has been awarded several prizes by both government and industry, as well as by academia, acknowledging its pioneering efforts and success despite an unfavourable macroeconomic environment (e.g. the period of hyperinflation and the current severe economic crisis prevailing in the country).

In 2002, the annual turnover of BioSidus reached US$45 million, which can be considered a good performance compared with other biotechnology companies across the world that received funding for a decade without generating any profits. About 75 researchers work in the company and collaborative links have been set up with research teams outside the company and even abroad. The current objective is to collaborate with other biotechnology companies in Latin America and the Caribbean in order to manufacture and commercialize products that are useful for the region.

On 4 April 2003, the chairman of BioSidus, Marcelo Argüelles, attended a seminar organized by the David Rockefeller Center for Latin American Studies and Harvard University on the theme "Joining the Revolution: Biotechnology as Business in Latin America". Argüelles referred to his company's current projects concerning the production of oral vaccines against cholera and typhoid, gene therapy to cure cancer and angiogenesis, as well as the development of transgenic farm animals to manufacture recombinant biopharmaceuticals (in addition to the production of EPO and interferons).

In January 2004, cloned calves Pampa Mansa II and Pampa Mansa III were born; they contained a gene coding for human growth hormone to be produced in their milk. The initial step in obtaining the first generation of clones involved taking fibroblastic cells from calf fetuses and then introducing the fetal cells' nucleus into the cytoplasm of enucleated ovocytes to produce embryos; the fetal cells were transformed to contain the gene for human growth hormone before their nuclei were transferred to enucleated ovocytes. The resulting embryos were transferred into Aberdeen Angus heifers. The procedure led to the birth of Pampa Mansa, which produced milk containing the human growth hormone in 2003. To create new clones, somatic cells were extracted from one of the transgenic Pampa Mansa's ears, then fused with enucleated ovocytes to produce embryos, which were transferred to surrogate mothers (*EFE News Services*, 8 February 2004).

BioSidus's spokeswoman, Vanesa Barraco, stated that the milk produced by just one cow could meet the demand of the entire nation for human growth hormone, and noted that 1,000 Argentine children required hormone therapy for hypopituitarism. Within about two years, the National Medicine, Food and Medical Technology Administration should approve the sale of the cloned animal-derived hormone. Argentina will be one of nine countries that have cloned genetically engineered cows since 2002, the year Pampa Mansa was born (*EFE News Services*, 8 February 2004).

BioSidus's Pharmaceutical Dairy Farm project was launched in 1999 and is being carried out by a multidisciplinary group. In 2004, US$4 million were spent out of what is expected to be a total outlay of US$6–7 million. The project enjoys the financial backing of the National Agency for Scientific and Technological Progress and Product Innovation. The project will obviously require a long-term investment of venture capital, but it is considered to be the best way to strengthen Argentina's capability in an increasingly competitive world and to position it among a select group of countries with advanced biotechnology capacity. BioSidus planned to export the growth hormone to Brazil and won an US$8 million contract from São Paulo State. National demand for the human growth hormone is estimated at around US$7 million, and global demand is approaching US$1 billion (*EFE News Services*, 8 February 2004).

Brazil's genomics programmes

At the University of São Paulo, Professor Fernando de Castro Reinach has been eager to link the public and private sectors of Brazilian science ever since he completed a PhD at Cornell University Medical School in the United States and a postdoctorate at the Medical Research Council's Laboratory of Molecular Biology in Cambridge, the United Kingdom. In 1990, with two other colleagues, he founded Genomic, one of the first Brazilian companies to perform DNA tests and one of the first to reach the market with a product. The company became involved in paternity searches and is currently one of the largest DNA-testing companies in Brazil (Greco, 2003).

In 1997, Reinach was among a group of seven Brazilian scientists to receive a grant from the Howard Hughes Medical Institute. During the same year, after receiving the approval of the scientific director of the State of São Paulo Research Foundation (FAPESP, Fundacão de Amparo a Pesquisa do Estado de São Paulo) – the country's third-biggest science and technology funding agency – Reinach put together a proposal to sequence the genome of *Xylella fastidiosa*, a bacterium that destroys Brazilian citrus worth US$100 million every year. In 1999, Reinach decided to launch another private venture, comDominio, an Internet hosting service that aimed to bring major sectors of the country online. Reinach worked closely with the former head of Brazil's central bank and a major backer of the new venture, which became the country's second-largest hosting service provider. Venture capitalist Paulo Henrique Oliveira Santos, president of Votorantim Ventures (a US$300 million company affiliated with Brazil's biggest industrial conglomerate), also invested in comDominio (Greco, 2003).

The *X. fastidiosa* project put Brazilian genomics on the world scientific map. High-throughput sequencing is a highly specialized activity, practised in a very limited number of laboratories in the industrialized countries. It is estimated that a dozen laboratories are contributing over half the total sequence data currently deposited in public databases, with another 50 or so accounting for the bulk of the rest. These laboratories are located in North America, the larger European countries, Australia and Japan. The latest entrant in this select club is Brazil, and more specifically the State of São Paulo (Adam, 2003).

This state has a law stating that 1 per cent of the tax revenue collected has to be allocated to the FAPESP. As São Paulo is the wealthiest state in Brazil, this amounts to a considerable budget (US$250 million in 1998). By law, the FAPESP cannot spend more than 5 per cent of its budget on administrative costs. The combination of ample funding and political independence gives the Foundation a lot of freedom to develop innovative scientific programmes (Adam, 2003).

In May 1997, the FAPESP decided that Brazil should not miss out on the scientific and economic opportunities to be derived from genomic sequencing, and should be able to produce its own data, analyse them and use the results for local scientific projects. In November 1997, after a call for applications, laboratories were selected to focus on sequencing the genome of *Xylella fastidiosa*, which causes Citrus Variegated Chlorosis. This choice also brought in additional funding from the citrus growers' association (FUNDECITRUS).

The concept of a single sequencing centre was rejected from the start. Instead, bids were sought from laboratories interested in participating in the project, and the selected laboratories received equipment (AB1370 sequencers), reagents and ample technical advice. In total, 30 laboratories were selected for the *Xylella* project, dispersed geographically throughout the State of São Paulo. In addition to the sequencing laboratories, the project steering committee designated a coordinator, Andrew Simpson, a molecular biologist at the Ludwig Institute for Cancer Research in São Paulo, and a bioinformatics centre. The bioinformatics group, located at the University of Campinas, was made responsible for all the data handling, from base calling to final assembly verification. The sequencing laboratories submitted trace files only, and were paid on the basis of the amount of non-vector, high-quality sequences that could be extracted from their data. Starting in March 1999, the *Xylella* sequencing project (estimated genome size: 2.7 million nucleotide pairs) was completed in January 2000, two months ahead of schedule, and the sequence was published in *Nature* on 13 July 2000. The budget allocated amounted to US$13 million, including US$250,000 from FUNDECITRUS.

This first experience in genomics led to the approval of a new programme in August 2000, carried out by a network of 65 laboratories – the Organization for Nucleotide Sequencing and Analysis (ONSA), or the Virtual Genomics Institute – involving 300 researchers. This new programme includes a human cancer genome project, funded by the FAPESP (US$10 million) and coordinated by Andrew Simpson; it is related to colon, stomach, head, neck and cervix cancers. The bids were sought from laboratories in April 1999 and the selection was made in June 1999. On 21 July 2000, the FAPESP announced the composition of 279,000 human expressed sequence tags (ESTs) – small pieces of DNA that allow genes to be located along chromosomes. The project goal is to analyse 1 million sequences and thus have a better understanding of the genes linked to cancers, particularly those affecting the head and neck, which for some reason are unusually common in Brazil. The new genome programme also comprises microbial genomes (*Xanthomonas axonopodis citri* – US$5 million; *Xylella fastidiosa*, which causes Pierce's disease in grapevines; *Leifsonia xyli*; *Leptospira*; and *Schistosoma mansoni*); bovine genomes (EST and functional genomics); and crop genomes (sugar-cane or SUCEST – US$6 million; coffee; and eucalyptus).

The ONSA's spin-offs are three companies:

- Alellyx-Applied Genomics, which is pushing forward genomic sequencing, and creating and using a large applied genomics platform to increase the productivity, competitiveness and quality of commercially important crops; its initial focus is on soybeans, oranges, grapevines, sugar-cane and eucalyptus.
- CanaVialis, which aims at developing new sugar-cane varieties; it was founded with US$7 million from Votorantim Ventures.
- Scylla Bioinformática, which offers solutions for both specialists and non-specialists dealing with biological data (genomic databases, genomics and proteomics, networking in bioinformatics).

Cuba

Cuba has spent a reported US$1 billion over the past 20 years building up its bio-industry (*The Economist*, 2003e). Research in genetic engineering for medical biotechnology had started by the early 1980s, the focus being alpha-interferon to treat cancer. At the beginning of the 1990s, the West Scientific Pole was created. This encompasses 53 institutions and scientific centres under the leadership of the Center for Genetic Engineering and Biotechnology, which was set up in July 1986. The basic objective of all these institutions is to participate in the country's system of public health care and to contribute to solving public health problems. Cuba has an immunization programme of its whole population with respect to 13 vaccines, an average life expectancy of 75 years and infant mortality of 6 per 1,000 live births. Since the approval in October 2003 of a vaccine against meningitis caused by *Haemophilus influenzae*, a hexavalent vaccine against meningitis B, diphtheria, tetanus, whooping cough, poliomyelitis and *H. influenzae* meningitis can now be used in the vaccination programmes in Cuba. The longer-term goal is to produce a heptavalent vaccine to immunize against seven diseases.

The Cuban strategy in medical and health-care biotechnology has the following characteristics:

- it is part of the national health-care system;
- it aims at solving the country's health problems;
- it is the result of a national endeavour, with proper human and funding resources;
- it is not yet opened to foreign investments.

The following drugs are produced via recombinant DNA technology: interferons, used as anti-viral and anti-proliferation drugs; anti-hepatitis B vaccine (which has almost eradicated the disease in the country); a cream containing recombinant epidermal growth factor (EGF), used in the treatment of burns and the cicatrization of wounds; recombinant streptokinase, used in the treatment of heart attack; and a wide range of diagnostics (e.g. to detect HIV in the blood of donors and patients). Currently research work is being carried out on the development of vaccines against dengue fever, cholera and HIV/AIDS. A vaccine against meningitis caused by meningococci B is, along with the anti-hepatitis B vaccine, a product that is well commercialized in Latin American countries (e.g. Brazil) and others (Iran). In 2000, a World Health Organization (WHO) inspection was followed by approval of Cuba's anti-hepatitis B vaccine for use in WHO-supported vaccination campaigns. Efforts have been made to master the commercialization process overseas better, and medical biotechnology products earn several hundred million dollars annually (*The Economist*, 2003e).

Many foreign firms might be deterred by the US embargo on the island and put off by the US Helms–Burton Act, which could shut them out of US markets for doing business in Cuba. Yet so-called "receptor companies", such as Canada's YM Bio-Sciences, which both develop and package Cuban products, have been busy agreeing joint venture licences with some of Havana's leading biotechnology centres (*The Economist*, 2003e). These companies are taking advantage of the high-quality and low-cost specialist humanpower in Cuba, where international production, management and regulation practices and norms concerning drugs are strictly followed.

Issues of intellectual property protection are also being resolved, so as to enable Cuba to penetrate the markets of developed countries. For example, Cuban officials state that their country enforces international protocols such as the World Trade Organization's Agreement on Trade-Related Aspects of International Property Rights (TRIPS). The Cuban state does own the intellectual property embodied in the products of the country's biotechnology research institutes or centres. But this, officials stress, is little different from the institutional ownership of patents in the United States by bodies such as university regents. The result in both cases can be good, affordable drugs (*The Economist*, 2003e).

On 16 July 2004, the California-based biotechnology company Cancer-Vax announced that the US government had authorized it to license three experimental anti-cancer drugs from Cuba, making an exception to the policy of tightly restricting trade with that country. CancerVax's officials stated that it was the first time an American biotechnology company had obtained permission to license a drug from Cuba. In 1999, SmithKlineBeecham, now known as GlaxoSmithKline, licensed a Cuban vaccine for meningitis B, which is currently being tested in clinical trials. The three drugs that CancerVax was going to test were first licensed to the Canadian company YM Bio-Sciences, which transferred those rights to CancerVax (Pollack, 2004b).

David Hale, chief executive of CancerVax, stated: "I think there are other product candidates and technology in Cuba that could be helpful not just to the American people, but people around the world." Cancer-Vax, a newly public company, did not yet have any drugs on the market. Its melanoma vaccine has been in development by an academic scientist for 40 years and was only currently in the final phase of clinical trials. The lead drug from the Cuban Center of Molecular Immunology (CIM) aimed to thwart epidermal growth factor in cancer cells; it has already been tested in small clinical trials outside the United States. In one trial, according to data presented in June 2004 at an American Society for Clinical Oncology meeting, patients with advanced lung cancer who received the drug lived longer than those who did not receive the treatment (Pollack, 2004b).

A spokesman for the US State Department, which helps rule on such licences, said that the exception had been made because of the life-saving potential of the experimental drugs from Cuba and the licence approval did not represent a relaxation of the trade policy. US members of Congress from both parties had sent letters to the Secretary of State urging that permission to license the drugs be granted on medical grounds.

Goldfield and Popkin, Washington lawyers hired by CancerVax to help win approval, stated that there had been no real opposition to the request. But they underlined that approval was more difficult to obtain than for SmithKlineBeecham's licence owing to the US administration's tougher policy toward Cuba (Pollack, 2004b).

CancerVax intended to test the drugs in clinical trials and bring them to market if they passed muster. The first drug, which had already shown some promise in small trials, could reach the market in 2008 or 2009, according to CancerVax's chief executive. The licensing deal called for CancerVax to pay US$6 million over the three-year period 2005–2007, during the developmental stage. If products reached the market, the company would pay as much as US$35 million more. A US condition of allowing the licence was that payments to Cuba during the developmental phase would be in goods such as foodstuffs or medical supplies, to avoid providing the Cuban government with currency. Any payments after drugs reached the market could be half in cash (Pollack, 2004b).

The Cuban Center of Molecular Immunology

The CIM, established in Havana, is a biotechnological institution devoted to basic research, product development and the production of mammalian cell culture products in compliance with good manufacturing practices (GMP). Employing more than 400 people, mostly scientists and engineers, the CIM has developed extensive competence in the field of monoclonal antibodies since 1980; its main research objective is the development of new products for the diagnosis and treatment of cancer and other diseases of the immune system. Current research projects focus on cancer immunotherapy, especially the development of molecular vaccines; they include antibody engineering, cellular engineering, regulation of the immune response and bio-informatics work. The CIM conducts clinical trials in diagnostic imaging and cancer therapy in specialized hospital units. The GMP production facility has been designed for maximum product protection. A positive-pressure air gradient system provides areas that meet class 100 to 10,000 specification. Hollow fibre and stirred tank fermentors are used in upstream production of industrial-scale mammalian cell cultures with an annual production capacity of several kilograms of recombinant proteins or monoclonal antibodies. Downstream production of injectable products is carried out through a rapid and highly automated technology. The CIM has quality control and assurance groups with highly qualified staff and the necessary equipment for sophisticated analytical and biological control of the production process and final products.

Since 1992, CIMAB S.A. – the exclusive representative for the marketing of the products and services of the CIM – has been selling the following products on the national market and abroad:

- A monoclonal antibody (anti-CD3) for the treatment and prophylaxis of renal transplant rejection (commercial name: ior t3 Muromonab CD3). It decreases the number of circulating T-lymphocytes by reacting and blocking the function of the 20 kd CD3 molecule on the membrane of human T-lymphocytes associated with the antigen recognition structure of T-cells and is essential for signal transduction. Ior t3 blocks all known T-cell functions and constitutes an excellent immunosuppressive compound.
- Recombinant human erythropoietin (commercial name: ior EPOCIM), with a molecular weight of 34 kd and 165 amino-acids, produced by mammalian cells into which the erythropoietin gene has been transfected. It is used for the treatment of anaemia (it is a glycoprotein produced in the kidney that stimulates the division and differentiation of red cells in bone marrow);
- Recombinant granulocyte colony-stimulating factor (GCSF) for the treatment of neutropenia in cancer patients (commercial name: ior LeukoCIM). It contains r-met-hu-GCSF, which regulates the production and release of functional neutrophils from the bone marrow and controls their proliferation, differentiation and other cellular functions.
- A humanized monoclonal antibody against epidermal growth factor receptor (EGF-R) for the treatment of tumours of epithelial origin (commercial name: CIMA her). It is a humanized immunoglobulin, isotype IgG, that binds the intracellular domain of the human EGF receptor; it has a potent anti-angiogenic effect inhibiting the *in vivo* and *in vitro* production of pro-angiogenic growth factors such as the Vascular Endothelial Growth Factor, and has important *in vivo* pro-apoptotic properties. It is used in combination with radiotherapy in the treatment of head and neck tumours, enhancing the anti-tumour response to 70 per cent, and no evidence of severe clinical toxicity was observed.
- Murine monoclonal antibodies for tumour imaging, used for *in vivo* diagnosis and monitoring of metastases and recurrences of several tumours of epithelial origin such as breast, lung, brain and colorectal tumours.
- Monoclonal antibodies targeted against tumours, including one in clinical trials – TheraCIM-CIMA her-h-R3, specific for the EGF-R, a molecule over-expressed on the surface of cancer cells, and used against head and neck cancers and gliomas – and three in pre-clinical development – C5Mab, an IgG1 humanized monoclonal antibody highly specific for colorectal tumour-associated antigen preferentially expressed on the surface of malignant colorectal and ovarian cells; 14F7Mab, an IGg1 monoclonal antibody highly specific for N-

glycosylated GM3 ganglioside, which recognizes human melanoma, colon and breast tissues; and 1E10 Mab, an anti-idiotype that identifies N-glycosyl-containing gangliosides and reacts with human melanoma and breast tumours.

• Therapeutic cancer vaccines and adjuvants. These include EGF-P64K, composed of recombinant EGF protein conjugated with a proprietary carrier protein (P64K), as well as a proper adjuvant, which is in clinical trials against uterine and cervix cancers; an NAcGM3/VSSP vaccine, based on the N-acetyl-GM3 ganglioside incorporated in a very small size proteoliposome (VSSP), and Montanide ISA 51 as adjuvant, which is in clinical trials against melanoma and breast cancer; an 1E10 anti-idiotype vaccine, based on anti-idiotypic Ab2 to anti-ganglioside antibodies, IgG, murine monoclonal antibody 1E10 combined with alum as adjuvant, which is in clinical trials against melanoma and breast cancer; a TGF vaccine, based on the fusion protein of the human recombinant Transforming Growth Factor and P64K protein combined with a proper adjuvant, which is in the pre-clinical development phase.

CIMAB is entrusted with the negotiation of research projects at different development stages, e.g. clinical trials in Canada and Germany of therapeutic monoclonal antibodies and vaccines against cancers. In addition, CIMAB has established commercial associations with more than 25 pharmaceutical companies worldwide for product distribution and licensing agreements. Two joint venture plants are being built in India and China to produce therapeutic monoclonal antibodies using the technology developed at the CIM; they will share the benefits equally. The current aim of the CIM and CIMAB is to play an increasing role in the developed world's drug market through licensing and joint ventures.

The Center for Genetic Engineering and Biotechnology is currently producing a therapeutic monoclonal antibody in plants in contained greenhouses. The aim is considerably to decrease the production costs as well as the usual drawbacks of using mammalian cells for this purpose. The CIM will check the biological activity and efficiency of the plant-derived monoclonal antibody. If successful, this process may lead to a joint venture with a multinational drug company to commercialize the product on the world market.

China

Chinese scientific research and development

Backed by good economic growth and spurred by the will to catch up with the West, China is investing heavily in science and technology. According to figures published in the 2003 *Overview of Science, Technology and Industry* by the Organisation for Economic Co-operation and Development (OECD), expenditure on R&D in China reached US$60 billion in 2001. China has therefore become the world's third science and technology power (by this standard) behind the United States and Japan. The budget of the Chinese Academy of Sciences – China's most prestigious scientific institution – more than doubled between 1995 and 2000 (Bobin, 2003).

With 743,000 researchers, China has the second-largest research population behind the United States, and this figure does not include the large number of Chinese people studying abroad. In 2000, more than 100,000 were studying in one of the OECD countries, particularly in the United States. In the latter, two-thirds of foreign students are of Asian origin, whereas Europeans account for only 17 per cent (Kahn, 2002).

Between 1992 and 1999, China rose from twelfth to eighth rank in the world with regard to the number of its scientific and technological publications. Its strengths lie in chemistry, physics, mathematics and engineering sciences, while its weaknesses are in basic and applied biology, ecology, medical research and astronomy. In 20 years, progress in R&D has been real, but we cannot yet speak of a "scientific power". No Nobel Prize or Fields Medal has yet been awarded to Chinese scientists, whereas Japan has received seven and three, respectively. In 1999, the US Patent and Trade Mark Office delivered only 90 patents to Chinese individuals, compared with 3,693 to Taiwanese. Despite China's large number of researchers, this represents only 11 scientists per 10,000 inhabitants, compared with a figure of 81.8 in the United States and 92.2 in Japan (Bobin, 2003).

The major concern about Chinese scientific R&D is the focus on the short term, to the detriment of the long-term approach. With the exception of some areas of excellence, which are privileged by the government, the state has abandoned automatic funding and has let down the principal actors in R&D. Consequently, laboratories and research institutes have to find their own means of funding and launch their own enterprises on the market. The Chinese Academy of Sciences has thus spawned 500 enterprises employing 40,000 persons. They are concentrated in the Beijing suburb of Zhongguancun, the capital's "Silicon Valley". The most renowned of these spin-offs is the informatics group Legend, which has 30 per cent of the Chinese personal computer market. However, this focus on marketable technologies also implies a lack of interest in basic research, which receives only 6 per cent of R&D expenses, whereas the percentage in the industrialized countries is 15–20 per cent (Bobin, 2003).

The obsession with the market and the immediate profits that might be derived has, according to Bobin (2003), deleterious implications for research ethics. At the September 2003 congress of the Association of Chinese Scientists, the biologist Zhou Chenglu vigorously denounced the "scientific corruption" that tends to thrive in China (e.g. the purchase of bogus degrees and the copying of PhD theses and publications) and is not sufficiently punished.

Efforts are being made to facilitate the return home of Chinese students or researchers living abroad. Of 580,000 students (in all disciplines) who left China between 1978 and 2002, only 150,000, i.e. about one-quarter, returned home. This proportion was one-third at the end of the 1990s. An increasing number of graduates are returning to China, lured by attractive posts in the technological start-ups or at the head of research laboratories whose staff is being rejuvenated. In 2001, the number returning to China increased by 34 per cent but, at the same time, the number of those leaving the country increased by 115 per cent. It is therefore too early to state that the brain drain has been stopped, when in addition there is an internal exodus to the multinational corporations established in China (Bobin, 2003).

The World Bank report *China and the Knowledge Economy, Seizing the 21st Century*, published in 2001, warned that, if innovation is not supported, China runs the risk of being sidelined technologically. This warning referred to the fact that China has missed an industrial revolution, although it nevertheless outpaced Europe in the Middle Ages, when it invented the compass, gunpowder, paper and printing. In China, it is a national sacred cause that this should not happen again (Bobin, 2003).

Investments in biotechnology

Between 1996 and 2000, the Chinese government invested over 1.5 billion yuan (US$180 million) in biotechnology; between 2000 and 2005, it planned to invest another 5 billion yuan. As a result, the Boston Consulting Group reckons that biotechnology is thriving in 300 publicly funded laboratories and around 50 start-up companies, mainly in and around Beijing, Shanghai and Shenzhen. One should note that in 2002 foreign direct investment reached US$52 billion for the whole of China, but almost one-third went to Guangdong province, especially the Pearl River delta, where Guangzhou (Canton) and Shenzhen are located (420,000 foreign corporations have invested US$450 billion in China over the past 20 years). The Chinese science ministry claims that as many as 20,000 researchers are working in the life sciences (Marti, 2003).

These figures look small when compared with the US$15.7 billion invested in R&D in 2001 by the US biotechnology industry, which was employing 191,000 people. But the speed with which China's bio-industry is growing causes both awe and anxiety in the outside world. There are currently more than 20 genetically engineered medicines approved for sale in China, earning 7.6 billion yuan in 2000. Yet, although they were made in China, these products are generally identical to developed-country inventions, introduced when China had little interest in intellectual property protection (*The Economist*, 2003a).

Genomics work

The director of the National Centre for Biotechnology Development admits that originality is an issue among Chinese scientists, but he hopes that the return of some of the 20,000 Chinese researchers working abroad will contribute to enhancing creativity. It is one of these returned scientists who heads the Beijing Genomics Institute (BGI). The BGI is the biggest not-for-profit genomics research institute in China. Established in July 1999, it has been growing rapidly with the support of the Chinese Academy of Sciences. The BGI has over 600 employees and two campuses (the main one in Beijing and another one in Hangzhou City, opened in January 2001), and enabled China to be the only developing country among the six countries in the International Human Genome Project Consortium. The BGI played a leading role in completing the sequence of the part of chromosome 3 assigned to it. In 2001, Sun Enterprise, a major software company, chose the BGI as a centre of excellence on the basis of its advanced work in genomics, alternative splicing algorithms and proteomics. The BGI team will use two Sun Enterprise supercomputers to study the rice and porcine genomes, among other projects (*The Economist*, 2002a).

The BGI and the Danish Porcine Genome Consortium launched the porcine genome-sequencing project in October 2000. The Danish consortium includes the Danish Institute of Animal Sciences, the Royal Veterinary and Agricultural University and representatives from Denmark's pig industry. The BGI will carry out the sequencing and sequence analysis work. In phase I, the project aims to identify the valuable genes to develop markers for physical and genetic mapping, and to provide the research tools for xenotransplantation in three years. During phase II, a working draft covering 90 per cent of the sequence and 95 per cent of the genes will be developed, bearing in mind that the pig genome is roughly the size of the human genome and is estimated to be made up of 3 billion nucleotide pairs (*The Economist*, 2002a).

The BGI also completed in 2000 the sequence of the genome of a rod-shaped, anaerobic, extremely thermophilic eubacterium isolated from freshwater hot springs in Tengchong in Yunnan province. *Thermoanaerobacter tengcongensis* has a circular genome of 2,689,445 nucleotide pairs, which was sequenced using a "whole-genome shotgun" approach. Its genome is very similar to that of *Bacillus halodurans*, a mesophilic eubacterium. Finally, the BGI is involved in the International HapMap Project, a five-country initiative launched in October 2002 to follow up the Human Genome Project with a large-scale study of human genetic variation and its relation to disease (*The Economist*, 2002a).

Medical and pharmaceutical biotechnology

The National Engineering Research Centre for Beijing Biochip Technology is headed by Cheng Jing, an engineer and molecular biologist trained in the United Kingdom and the United States. The Centre has already spun out some of its technology to Chinese and US start-ups. It had two diagnostic chips for infectious disease and tissue transplantation in trials in Beijing hospitals, and it was spearheading a drive to link most of China's biochip expertise under one roof in a Beijing science park by 2003–2004 (*The Economist*, 2002a).

Another advanced medical biotechnology hotspot is the Chinese National Human Genome Centre in Shanghai, whose focus is to study the genetics of diseases that particularly affect the Chinese population, such as hepatocellular carcinoma. In 2003, the number of people in China suffering from hepatitis B was estimated at 130 million, and the annual death toll at 250,000. Schistosomiasis (bilharziasis) affected nearly 1 million people. There were 1.3 million new cases of tuberculosis annually on average and 250,000 deaths. In 2001, HIV/AIDS killed 30,000 people annually out of 1–1.5 million sufferers. In 2002, measles – a highly contagious disease – killed 7,000 people among 58,341 sufferers. These figures from the World Health Organization, the World Bank, the Chinese Ministry of Health and the Institute of Development Studies underline that China is reeling from an onslaught of communicable diseases; hence the efforts being made by China to develop preventive and therapeutic measures, including medical biotechnology.

In Hong Kong, the Biotechnology Research Institute is screening compounds isolated from traditional Chinese remedies to check if they have any effect on receptors known to be involved in neurodegenerative diseases. The Shanghai Traditional Chinese Medicine Innovation Centre has been working with PhytoCeutica, an American biotechnology company, to set up a database of 9,000 traditional herbs and 150,000 recipes (*The Economist*, 2002a).

Stem cell research in China is mostly focused on adult cells, and half a dozen stem cell banks have already been established. At Shanghai Second Medical University, work is being carried out on generating stem cells by transferring nuclei from human skin cells into rabbit ovocytes.

The aim of this research is to understand the early stages of cellular re-programming better; this requires thousands of ovocytes, which are un-available from human sources.

Cooperation

Most Chinese medical biotechnology is funded by the government, al-though some investors from Taiwan, Singapore and Hong Kong are in-creasingly interested. Venture capital groups from the United States and Europe are generally waiting until they can be assured of a way to recoup their investment. Another shortcoming is intellectual property protec-tion. Although China strengthened its patent laws for drugs and other biotechnology-derived products in 2001, it is still difficult and expensive to obtain a patent and to exercise it owing to weak enforcement mecha-nisms (*The Economist*, 2002a).

Cooperation with developed countries in the life sciences and medical biotechnology (for example with the European Commission) is growing and could help China overcome some of the shortcomings that are hin-dering its bio-industry. For instance, a general cooperation agreement concluded with France specifies the functioning of joint research carried out in the life sciences and genomics in Shanghai. In this agreement, re-spect for ethical principles is considered a top priority, followed by ways of sharing industrial property and the mechanisms for valorizing research results. The French partners include the National Scientific Research Centre, the National Institute for Health and Medical Research and the Institut Pasteur; the Chinese associates are the Rui Jin Hospital, the Chi-nese National Human Genome Centre and Shanghai Second Medical University (Kahn, 2002).

On 29 January 2004, at the Institut Pasteur in Paris, the vice-president of the Chinese Academy of Sciences, Chen Zhu, and the director-general of the Institut Pasteur, Philippe Kourilsky, announced the creation in Shanghai of the first Chinese Institut Pasteur (the former Institut Pasteur of China, also based in Shanghai, was closed down in 1950). This is not an "outpost" of the French institution but a fully Chinese organization in which a group of French researchers will be working in close collabora-tion with their Chinese colleagues. The new institution, devoted to teach-ing and research, will be almost entirely funded by the government of China and will have, like its French counterpart, the status of a "private foundation" – an innovation in the scientific arena. Some 250 members will constitute the staff of that institution, considered to be an outstand-ing example of the close cooperation existing between China and France, which is based "on the principles of equity, sincerity and mutual inter-ests" (Nau, 2004a).

India

India spent US$19 billion on R&D in 2001, and is among the top 10 countries in the world (Kahn, 2002). In 2004, 10 biotechnology-derived drugs were being marketed, four industrial units were manufacturing recombinant anti-hepatitis vaccines locally, and indigenously produced recombinant erythropoietin and granulocyte colony-stimulating factor were also on the market. There are several recombinant drugs and vaccines in advanced stages of production.

India has become a world leader in the production of generic drugs. It is nowadays an attractive destination for contract research organizations (CROs), businesses that run trials for pharmaceutical groups. These clinical trials – the approval process for any new pharmaceutical – are time-consuming, expensive and ethically tricky; the task involves recruiting hundreds, often thousands, of sick people to volunteer for the testing of experimental medicines with unknown side-effects. The aim of carrying out clinical trials in India is to reduce the time and funds needed to turn new molecules into marketable drugs, a process that can take up to 20 years and cost US$800 million per drug developed (Marcelo, 2003).

According to the consultancy firm McKinsey, the overall cost advantage in bringing a drug to market by leveraging India aggressively could be as high as US$200 million, because of the availability of large patient populations, access to highly educated talent and lower-cost operations. These developments occur when pharmaceutical companies are beginning to consider transferring parts of their research operations to India. India is attractive because of its many scientists and the fact that it is implementing tougher patent protection. Some executives believe India could become as prominent in pharmaceuticals as it is in information technology (Marcelo, 2003).

In clinical trials, India, unlike the United States, offers a huge pool of what the industry calls "treatment-naive" patients – those who have not been tested with rival drugs. A larger pool of such people may lead to faster patient enrolment in trials and thus to more rapid drug development. In India, there are about 30 million people with heart diseases, 25 million with type-2 diabetes and 10 million with psychiatric disorders. These widespread supposedly "rich world" diseases are considered an important target for companies looking to test drugs destined for "western" consumers (Marcelo, 2003).

The world's largest CRO, USA-based Quintiles, began operations in India in 1997 and has recruited 6,400 patients for clinical trials in areas such as psychiatry, infectious diseases and oncology. In 2003, a dozen CROs had set up office in India, up from three in 2001. Mike Ryan, business development manager of CRO Pharmanet, which had been in India for about a year, stated that one of the country's attractions is that patients hold physicians in high esteem. As a result, patient compliance in trials is high – in contrast to the United States, where subjects often drop out to seek second opinions. Cathy White, chief executive of Neeman Medical International, a USA-based subsidiary of India's Max Health Care group, stated that companies could save 20–30 per cent in drug development costs by outsourcing to India. Most of these savings come from hiring clinical researchers, nurses and information technology staff at less than one-third of "western" wages (Marcelo, 2003).

Another factor underpinning the shift to drug testing to India is the recent change in medical research rules. In 2003, the Indian health authorities adopted guidelines on "good clinical practice" in line with global norms. Nevertheless, some inside the pharmaceutical industry argue that, if patients are illiterate, there are serious ethical issues over their consent in a drug trial. Allan Weinstein, vice-president of clinical research and regulatory affairs with Eli Lilly & Co., has stated that "India should not be a place to go just because there are a lot of fresh patients". There must be a likelihood that patients involved in a clinical trial will benefit from the drug (Marcelo, 2003).

Singapore

Although Singapore is not considered a developing country, its development in the area of biotechnology has features in common with the way it occurred in the technologically advanced developing countries (e.g. China). Singapore began promoting biotechnology in the early 1980s, attracting Glaxo in 1982. During the following decade, Singapore pushed research and development, setting up a Bioprocessing Technology Centre Incubator Unit for start-ups, with fully equipped laboratories, in 1997. Yet Singapore spent and is still spending less as a proportion of its gross domestic product on R&D than Japan, South Korea or Taiwan. Moreover, investors are shying away from an industry where products take at least a decade to develop. On the other hand, increased competition is coming from less developed countries such as China, India and Malaysia, which are building a bio-industry of their own. Cheap labour in China is drawing jobs away and the government has warned that the unemployment rate, currently 4.5 per cent, was likely to climb to 5.5 per cent in 2003, its highest rate since 1987, with the economy likely to grow at no more than 1 per cent (Arnold, 2003).

Faced with declining returns in electronics, the industry that helped move Singapore into the ranks of the world's wealthiest nations, the government is throwing its administrative power – and at least US$2.3 billion in investments, grants and other incentives – behind an endeavour to become an integrated biotechnology hub. Singapore needs to find a new niche for its economic and social development (Arnold, 2003).

The biotechnology initiative has attracted other big-name manufacturers and research talent. Singapore has had the most success in attracting drug companies with tax holidays and other incentives. Among those with factories there are GlaxoSmithKline, Wyeth, Merck & Co., Schering-Plough and Pfizer. On 18 October 2000, Pfizer announced future investment of US$340.6 million in Singapore in order to build its first plant in Asia; this unit has been producing medical ingredients for the manufacture of drugs since 2004 with a staff of 250.

In fact, the Singapore government has sought to attract global drug-makers with a five-year, US$1.8 billion programme that includes research funding, start-up capital, tax breaks and new facilities. That is why Novartis AG, the world's fifth-biggest drug-maker, has decided to join several other rivals in order to make Singapore the site for what was expected to be its biggest drug-manufacturing plant in Asia. Daniel Vasella, Novartis AG's chairman, stated that the company had selected Singapore because of its research facilities and political stability. Novartis AG's new research institute, which formally opened on 5 July 2004, will focus on developing drugs against dengue fever and other tropical diseases. The company has agreed that its institute will provide research training to local scientists. Singaporeans were expected to fill a quarter of staff posts at the research unit. Eli Lilly and Viacell also opened research institutes, the costs of which were partly supported by government grants. Singapore has also opened a biopharmaceutical facility with the goal of developing drugs that could be provided on a contract basis to other companies (Burton, 2004).

Pharmaceutical production swelled about 50 per cent in 2002, to US$5.56 billion, but this industry is less labour intensive than the electronics industry. To encourage companies to do more than make drugs, the Economic Development Board offers to pay up to 30 per cent of the cost of building R&D facilities (Arnold, 2003).

In 2000, Singapore declared biotechnology as the "fourth pillar" of its economy and spent approximately US$570 million to set up three new biotechnology research institutes. By the end of August 2003, Singapore was putting the finishing touches to a US$286 million Biopolis medical R&D complex, built by JTC Corp., the government's industrial park operator. Biopolis comprises a huge underground vivarium to house the rodents needed for the research and is surrounded by a high-technology campus of about 200 ha, complete with condominiums, schools and wireless Internet access (Arnold, 2003).

Singapore is close to the Equator and has a lot to offer for the study of tropical diseases endemic to the region, such as malaria, and its own population is affected by illnesses of affluence such as cancer and heart disease. Its advanced telecommunications infrastructure and plentiful computing resources are another attraction – hence the increasing use of bio-informatics in drug discovery. For stem cell researchers, Singapore offers one of the world's most liberal legal environments. It allows stem cells to be taken from aborted fetuses and human embryos to be cloned and kept for up to 14 days to produce stem cells. In 2002, these welcoming conditions attracted Alan Colman, the scientist who helped clone Dolly the sheep in 1996 and who later moved to ES Cell International, a joint venture between Australian investors and Singapore's Economic Development Board, in order to pursue his medical research (Arnold, 2003).

7

Social acceptance of medical and pharmaceutical biotechnology

Social acceptance

All surveys and enquiries about the public perception or social acceptance of biotechnology (in both developed and developing countries) show undisputed support for medical and pharmaceutical biotechnology, whose benefits are acknowledged by a high majority of respondents and interviewees. For most people, health care is a top priority and anything that may improve it is more than welcome. Medical biotechnology obviously has a contribution to make in terms of more accurate and faster diagnosis of diseases and identification of pathogens, of prevention (safer, more effective and eventually cheaper new vaccines) and of therapy (new drugs, increasingly derived from genomics).

The issue of the increasing resistance of microbial pathogens and parasites to drugs does not deter patients from taking the relevant drugs. Combating this resistance is part of pharmaceutical research and development (R&D), which has to discover more efficient drugs, as well as of sanitary measures in hospitals to eliminate or mitigate nosocomial diseases, and of education of patients regarding drug posology.

The overall social acceptance of medical and pharmaceutical biotechnology is also due to the reliability of the drug approval process and the credibility of the relevant agencies (e.g. the US Food and Drug Administration – FDA). Biovigilance is considered a good safeguard against an eventual health hazard, because it entails the immediate withdrawal of any suspect drug. Social acceptance issues in medical biotechnology arise more as ethical issues when it comes to using genomics information to discriminate against people (in terms of recruitment, life insurance, etc.), because people are not equal with respect to their vulnerability to diseases and other stresses; to the screening of human embryos before implanting them in the mother or a surrogate; to xenotransplants and cloning.

Bioethics

The word "bioethics" was coined by V. R. Potter (1970), to mean "the science for survival" that "would attempt to generate wisdom, the knowledge of how to use knowledge for social good from a realistic knowledge of man's biological nature and of the biological world". A generalized and simple definition was proposed by Macer (1998) as a "love of life" involving analysis of the benefits and risks arising out of the moral choices affecting living organisms for the good of individuals, the environment and society.

Bioethics does not denote a particular field of human enquiry but is at the interface between ethics and the life sciences, emerging as a new area and concern in the face of great scientific and technological changes, connecting medicine, biology and environmental sciences with the social and human sciences such as philosophy, theology, literature, law and public policies (Bhardwaj, 2003). The four fundamental principles of bioethics are: beneficence, described as the practice of good deeds (doing good is beneficence); non-maleficence, which emphasizes obligations not to inflict harm; autonomy, which is the guiding principle for recognition of the human capacity for self-determination and independence in decision-making; and justice, based on the conception of fair treatment and equity through reasonable resolution of disputes (Bhardwaj, 2003).

Biocentric thinking in bioethics focuses on each individual organism. It may include the role played by each organism in the ecosystem, and it emphasizes the value of each life equally in decision-making or the consequences for an organism. Ecocentric thinking focuses on the ecosystem as a complete dynamic system and on the interrelationships between different entities of the system. An ecocentric system does not identify each individual life separately but takes a holistic and altruistic approach to the ecosystem, rather than looking at the impact of one species on the whole system. Anthropocentric thinking focuses on human beings and their interaction with nature. It is sometimes criticized by environmentalists and animal rights activists as based on a "self-love" approach that does not give equal and due importance to other living beings in the ecosystem (Bhardwaj, 2003).

Descriptive bioethics concerns the way people view life, and their moral interactions with and responsibilities towards the living organisms in their life. Prescriptive bioethics tells others what is ethically good or bad, or what principles are most important in making such decisions. It may also say that something or someone has rights, and that others have duties towards them. Interactive bioethics involves discussion and debate between people, groups within society, and communities about descriptive and prescriptive bioethics. It increases communication and dialogue within societies to clarify doubts and tries to develop a universal acceptability of things (Bhardwaj, 2003).

The components in the ethical debates about biotechnology are shaped by the way genetic engineering is viewed. Ethical choices are also shaped by individual reflection or by a holistic approach. Environmental non-governmental organizations oppose the use of genetic engineering on the basis of a more biocentric view; their goal is to protect the environment at any cost, which is sometimes also considered radical given the other demands of society, although it strongly favours the ethical principle of doing no harm. Some governments try to meet the needs of people and conserving the environment by taking a more balanced approach – the sustainable use of technology without causing undue harm to the environment. This may be socially, environmentally and, obviously, politically important. Anthropocentric and ecocentric views can be considered to be based on the ethical principles of beneficence and justice. The profit-oriented approach of the private sector – using the environment for economic gain – is based on the ethical principle of autonomy; the ultimate goal of the private sector is to produce the maximum benefit and economic returns for investment, which the private sector usually defends in the name of social development. There is no philosophical basis for complete abstinence from biotechnology, and bioethical principles advocate critical analysis of the benefits and risks of technologies only so that any unintentional harm (morally, theologically, socially and scientifically) can be minimized (Bhardwaj, 2003).

Pre-implantation genetic diagnosis

After 25 years of staggering advances in reproductive medicine – first, "test-tube babies" obtained after *in vitro* fertilization (IVF) and embryo transfer to the mother, then donor ovocytes and surrogate mothers – science and technology can help couples to have the kind of babies they want. For instance, by allowing the choice of gender, pre-implantation genetic diagnosis (PGD) may obliterate one of the mysteries of procreation. Couples from a wide range of cultures, nationalities and religions all share a powerful drive to have children of their own genetic stripe.

Physicians in a few countries, including India, South Korea, Israel, Italy and the United States, have begun to meet the expectations of this international clientele.

At the Genetics and IVF Institute in Fairfax, Virginia, an FDA clinical trial of a sperm-sorting technology called MicroSort was under way by early 2004. The clinic has recruited hundreds of couples, and more than 400 babies out of the 750 needed for the trial have been born. PGD, though, is by far the most provocative gender-selection technique. Some clinics offer the procedure for couples already undergoing fertility treatment, but a small number are beginning to provide the option for otherwise healthy couples. The Fertility Institutes in Los Angeles, headed by Jeffrey Steinberg, which has an office in Mexico, had performed its hundredth PGD sex-selection procedure by early 2004; one-third of its clients had travelled from Hong Kong, Egypt, Germany and other countries.

Some countries are beginning to clamp down even on less controversial fertility procedures, which have been taken to morally questionable lengths. For instance, British and Italian medical boards have questioned Severino Antinori's ethics; in 1994, he helped a 62-year-old woman become the oldest to give birth, and even claimed to be trying to clone a human. The Vatican, which calls the Italian gynaecologist's work "horrible and grotesque", pressured the Italian parliament to pass new laws in December 2003 outlawing surrogate parenthood and IVF for elderly couples. This prompted a minor panic at Italy's 2,500 fertility clinics, which will have to scale down the range of services they offer. In particular, gender selection thanks to PGD may make matters worse by galvanizing opponents of assisted-reproduction treatments.

Already, many couples are forced to travel far and wide for access to the latest procedures. In the near future, they almost certainly will have fewer places to go. Even China, considered a haven for far-out medical research, has actually been more diligent in its legislation than countries such as Italy so far. In 2002, surrogacy and payments to ovocyte donors were banned, and the number of IVF cycles a clinic can perform was restricted. In the autumn of 2003, the authorities moved to ban advertising of infertility treatments and restricted the number of fertilized eggs that can be implanted to two for each woman under 35 and to three for each woman over 35. As in India, where it is illegal to use ultrasound or amniocentesis to determine the sex of a baby (for fear that female fetuses will then be aborted), it is not so much the law that is the problem as social attitudes. The most dangerous gender-selection approach – one that is skewing sex ratios in Asia's largest populations – is simple female infanticide.

Fertility has nevertheless become a fast-growing "industry" in India and in South Korea as well. Israel's 30 IVF clinics are also very active, and the country produces more scientific papers per capita on fertility than any other nation. The Assuta Hospital in Tel Aviv performs almost 4,000 cycles of IVF each year. Because treatments are subsidized by the state and labour costs are quite low, the clinics charge overseas patients only about US$3,000 a cycle for IVF – a quarter of the charge in the United States. In 2003, the Israeli government tried to cut back on IVF treatments (to one child per family from two), but then backed off in the face of strong opposition.

However, the benefits from pre-implantation genetic diagnosis can be illustrated by the story of Molly Nash, who was born with a rare disorder, Franconi's anaemia, which causes bone marrow cells to fail. Molly needed new cells from a donor who is an almost exact genetic map. With the help of PGD (the test can be performed in 24 hours, with time to spare for implanting the embryo into the womb), Molly's parents conceived their son Adam, who successfully donated umbilical-cord blood to save his sister's life. So far, the PGD has been used largely, as in Molly's case, in laudable efforts to avoid passing on single-gene inherited diseases. But PGD is transforming reproductive medicine by giving parents unprecedented control over what genes their offspring will have. It makes some people concerned, because it also gives physicians a rudimentary ability to manipulate traits – the morally reprehensible practice of eugenics. The fear is that, as other aspects of reproductive technology improve, PGD may be misused.

Are the benefits worth the risk? Molly's parents and many others think so. Clinics in London, Chicago, Tel Aviv and Brussels have began to offer PGD, and dozens of obstetricians have sent patients to the laboratory of Mark Hughes – the molecular biologist who has worked for 10 years building and perfecting PGD. The process starts with the arrival of tiny plastic tubes packed in ice. They contain a single human stem cell plucked a few hours before from a three-day-old embryo. The cells come from fertility clinics, where would-be parents have their eggs harvested, fertilized and grown in Petri dishes. By day three, a human egg cell has managed to divide, on average, into only six stem cells. To find out if it carries the genes for Tay-Sachs or cystic fibrosis or sickle-cell anaemia, the laboratory's researchers and technicians copy the sample cell DNA and analyse it. The technique has aroused controversy in the United States simply because it involves embryos. In 1997, Hughes was accused of using federal funds for embryo work. He lost his funding and resigned from Georgetown University. He moved to Detroit and set up the Center for Molecular Medicine and Genetics at Wayne State University.

The spectre of eugenics nevertheless looms on the horizon. At least one clinic in the United States is currently offering PGD services that allow parents to select the gender of their child, and more will probably follow. Hughes does not condone the practice. He admits that he has struggled with the question of whether PGD specialists are unwittingly turning children such as Adam Nash, selected to provide a transplant for his sister, into commodities; he has never managed to answer the question unequivocally. He adds: "what is wrong with our having a child we are going to love very much, but who also has the miraculous power to save our other child's life?" It is not an easy question to answer.

According to the Reproductive Genetics Institute in Chicago, US biologists have used genetic tests to help five couples to give birth to babies whose bone marrow will be grafted to a sibling suffering from acute leukaemias or rare blood diseases. All these diseases require the grafting of adult stem cells that are immunologically compatible with the patient's immune system. The blood extracted from the umbilical cord of one of these babies has enabled a therapeutic treatment to be undertaken on a sibling, and another child was awaiting a transplant (*Journal of the American Medical Association*, 4 May 2004). Three other children, whose parents had relied on *in vitro* fertilization and PGD to select an embryo for therapeutic purposes, were in remission.

On 10 December 2003, within the framework of a revised law on bioethics, French parliamentarians authorized the experimental use of PGD for parents with a child affected by an incurable lethal disease. These parents will be given access to medically assisted procreation aimed at selecting an embryo not affected by the illness (Benkimoun, 2003).

Another amendment adopted by the French National Assembly allows a procedure that consists of collecting stem cells from the second child's umbilical cord and using them therapeutically to cure the elder child. This means choosing an embryo whose genetic make-up fits that of the ill child and is therefore capable of providing a cure. The parliamentarian who proposed this amendment to the law described the child to be born as a "double-hope child" rather than a "medicine child", the name used previously. Another parliamentarian who supported the amendment wondered whether "one may refuse a couple the possibility that a future child may also provide, under strictly defined conditions, the means to extend the life expectancy of his/her elder sibling affected by a lethal disease" (Benkimoun, 2003, p. 9).

In this respect, it is worth mentioning the views of James Watson, the co-discoverer of DNA structure in 1953, on the humanitarian approach to using genetics: "it is part of human nature for people to want to enhance themselves. When someone is good-looking or bright there is a tendency not to care about those who are not. Thirty years ago cosmetic surgery was almost amoral; now hardly a politician can survive without it. To want your children to have a good throw of the genetic dice is extremely natural". Watson's vision for genetics therefore extends far beyond curing disease, and his willingness to discuss the ethical vanishing point of genetics has sometimes obscured a genuinely humanitarian approach to limiting human suffering (Swann, 2004, p. W3).

Watson stresses that genetics should be put in the hands of the user – more often than not, women. "These kinds of issues should not be decided by a group of government-appointed wise men. We should leave it up to women and let them make their own choices ... There is far too much regulation. If nobody is hurt, then what is going too far?" Watson's impatience to make genetics practical was heightened by the illness of his son, who suffers from a kind of autism. It was around the time of his son's diagnosis in 1986 that Watson was appointed chief cartographer in charge of mapping the human genome. For Watson, his son – now hospitalized – is a symbol of the genetic injustice that could be alleviated by progress in molecular genetics. "So far," he states, "the biggest practical impact of genetics has been in paternity suits and forensics. But they have found two genes for autism, so one day autistic children may not be born" (Swann, 2004, p. W3).

However laudable his motives, Watson's combination of impatience and unorthodox views generated negative reactions among fellow scientists. Watson, who was 75 in 2004, nevertheless advocates an understanding of genetics as an antidote to some of the self-delusions to which humans are prone. The fact that we now understand that humans are animals, he stated, should help us overcome some of the guilt that accompanies many of our desires. The discovery of DNA was indeed the final element in the Copernican revolution that displaced humankind from the centre of the universe. "Human beings were even more mysterious before 1953," he stated (Swann, 2004, p. W3).

Stem cells

The use of stem cells for regenerative medicine

The use of stem cells derived from embryonic cells is also a thorny issue. Embryonic stem cells have medical promise because they have the capacity to become any one of the more than 200 cell types making up the human body. Geron, the most advanced of the firms that are studying these cells, has worked out how to lead embryonic stem cells to turn into seven different types of normal cell line, which may be used to repair damaged tissue (heart, muscle, pancreas, bone, brain in Parkinson's disease, spinal injury, and liver).

Liver transplant operations have become almost routine, and finding suitable donors remains the hard part. Each year thousands of patients in the United States suffering from cirrhosis, hepatitis and other liver ailments die while on waiting lists. Artificial livers are not likely to fill the gap either. The liver is almost as complex as the brain; it handles a large number of physiological functions, from detoxifying the blood to turning food into the nutrients and chemicals the cells need to function and survive. At the Department of Experimental Surgery at Berlin's Charité Hospital, scientists have developed a small bioreactor containing a matrix of hundreds of membranes, within which they have coaxed human adult liver stem cells to grow into complex living tissue remarkably like a healthy liver. When the researchers feed a patient's blood through the bioreactor, the cultured liver cells take over all the normal, healthy functions of the patient's own diseased organ. The bioreactors are being used in clinics in Berlin and Barcelona to save the lives of patients whose own livers have stopped functioning but whose donor organs have not yet arrived. Jörg Gerlach, who heads the Charité Hospital's team, hopes to use the liver's regenerative capacity to make many transplants unnecessary in the future by hooking patients up to the reactor so their own livers can take time off and recuperate. Gerlach's next project, which is under way at the University of Pittsburgh, is to use the tissue culture techniques developed for the bioreactor to induce the body to grow new liver tissue on its own.

Type-1 diabetes affects 5 million people worldwide, but type-2 diabetes affects 150 million people. The key distinction is that type-2 diabetes is characterized by the inability to utilize the insulin produced by the body, whereas type-1 diabetes is an autoimmune disease in which the body's immune system attacks the islet cells in the pancreas that produce insulin.

In the technique developed by physician James Shapiro in Canada, known as the Edmonton Protocol, islet cells from the pancreas are implanted into the liver, where they develop a blood supply and begin producing insulin. Previous attempts to transplant islets were successful in only 8 per cent of cases, but with this new method 89 per cent of patients were still producing insulin after three years. However, the benefits of renewed insulin production have to be offset against the potential problems arising from taking immunosuppressant drugs. An older technique is to transplant the whole pancreas, often with a kidney, in people who are suffering renal failure or life-threatening hypoglycemic episodes of which they are unaware. The combined transplant has a very good success rate, with 80 per cent of patients alive after 10 years, but it is not as frequently performed as physicians would like because of the shortage of donor organs (Gorman and Noble, 2004).

Researchers are exploring such possibilities as turning embryonic stem cells into unlimited numbers of insulin-producing cells, or using adult stem cells obtained from the patient's bone marrow or liver or pancreatic tissue to do the same. Others are trying to collect islet cells from genetically modified pigs and even from fish that have been engineered to produce human insulin. Researchers believe that they have cured type-1 diabetes in mice, but they have yet to translate that success to humans, although they hope to begin trials soon (Gorman and Noble, 2004).

In May 2004, in the weekly British medical journal *The Lancet*, a group of US biologists led by Edward W. Scott and Dennis A. Steindler, in charge of the programme on stem cells and regenerative medicine at the University of Florida, published the results of their recent work on how bone marrow stem cells used for therapeutic transplants evolve in the human body. The US researchers have been trying to confirm an earlier preliminary observation, published in 2000, concerning the migration of some of these stem cells toward the brain of the person receiving the transplant (Nau, 2004e). They analysed samples from the brain of three women who had died from leukaemia after having received a bone marrow graft. These women were 45, 39 and 29 years old, respectively, when they died. In all three cases, the donor was the brother of the patient. The researchers found neurons in some brain regions that contained a Y chromosome. The presence of this chromosome had a frequency of 1–2 per cent in the interstitial brain tissue or glia. In the third case, the US biologists also found neurons and astrocytes bearing the Y chromosome. These results provide convincing evidence of the migration of haematopoietic stem cells into brain tissues (Nau, 2004e).

The publication of these results occurred at a time of hot debate between those who support clinical trials using embryonic stem cells to combat some degenerative diseases and those who believe that, for ethical reasons, only adult stem cells should be used. However, new studies are challenging earlier data and raising questions about how malleable and powerful adult stem cells really are. Leonard Zon, president of the International Society for Stem Cell Research, stated: "people are starting to realize that the science of plasticity is not all there" (Kalb, 2004).

Scientists know most about adult blood stem cells, which have been used for decades in bone marrow transplants for patients with cancers or blood diseases. That success prompted researchers to wonder whether adult blood stem cells could have the same ability as their embryonic counterparts. Initially, Markus Grompe, of Oregon Health and Science University, thought this would be the case. In 2000, he reported that adult blood stem cells were able to turn into liver cells in mice. Other studies reported that the cells could become neurons and heart muscle. In 2002, however, several scientists proposed a new theory: adult bone marrow cells were not actually becoming new tissue types; they were fusing with existing cells. These findings persuaded Grompe, a Roman Catholic who does not study embryonic stem cells on religious grounds, to re-assess his data and admit that blood cells were indeed merging with liver cells (Kalb, 2004).

In 2001, the news that bone marrow stem cells turned into heart muscle after being injected into the damaged hearts of mice aroused great hope, and even led to human trials. Piero Anversa, of New York Medical College, worked on one of the original studies and stood by the research. But other scientists were sceptical, and in 2004 two groups reported that they could not reproduce the earlier findings. Scientists believe therefore that the greatest potential of adult stem cells may be in regenerating the organs they come from, and they are actively searching the stockpiles of cells in different parts of the body. For years, researchers hoped that adult stem cells in the pancreas could be coaxed into becoming insulin-producing beta cells, which people with type-1 diabetes lack. But, in May 2004, Harvard University's Doug Melton dashed the hopes of many when he reported that he could find no adult stem cells in the pancreas at all. Although the study did not rule out their existence, the conclusion was clear to Melton: "if you want to make more beta cells, the place to look to is embryonic stem cells" (Kalb, 2004).

Opponents of the research on embryonic stem cells were looking to a unique group of adult stem cells isolated by Catherine Verfaillie at the University of Minnesota's Institute of Stem Cells. She had reported that the cells appeared to have some of the transformative capacity of embryonic stem cells. However, Verfaillie stated that the cells' transformative ability compared with that of embryonic stem cells remained to be seen. In fact, stem cell science is constantly evolving: even with all the recent challenges to plasticity, new studies of adult stem cells continue to report intriguing successes. As a result, scientists insist research must proceed along both pathways – with both adult and embryonic stem cells (Kalb, 2004).

Stem cells and cancer

At the annual meeting of the American Association for Cancer Research, which took place in Orlando, Florida, at the end of March 2004, Michael Clarke of the University of Michigan reported that a small population of slow-growing cells in tumours – cancerous stem cells – may be responsible not only for the recurrence of tumours but for the original cancers as well. To test his hypothesis, Clarke used a technique that distinguishes between fast- and slow-growing cancer cells. The technique consists of mixing the cells to be sorted with antibodies that have fluorescent tags attached to them. Each type of antibody will adhere to only one sort of protein found on a cell's surface – and each sort is given a different tag, so that it glows a different colour. Cells that have different combinations of surface proteins thus glow in different colours. They can accordingly be sorted by a device that works like an ink-jet printer. Cells are sprayed out in individual, electrically charged droplets, recognized on the basis of their colour, and sent to their destination test tubes by manipulating the trajectories of the droplets with an electric field (*The Economist*, 2004c).

The US researcher was able to separate tumour cells from breast cancer biopsies into fast-growing cells, which make up about 75 per cent of a tumour, and slow-growing ones, which comprise the remainder. When injected into mice, the fast-growing cells were unable to generate a new tumour, even when 50,000 cells were injected. By contrast, as few as 200 of the slow-growing cells gave rise to a new tumour. Even more important, the new tumour formed by these slow-growing cells looked exactly like the original tumour removed from the patient. Furthermore, when the tumour was ground and its cells were sorted by marker proteins, the researcher found both fast- and slow-growing populations, in similar proportions to those found in the tumour (*The Economist*, 2004c).

The simplest interpretation of these data is that the slow-growing cells represent a stem cell population for the tumour. If this is the case, as Clarke believes, it means that, even if other cells in a tumour are eliminated by chemotherapy or another treatment, the stem cells will persist. Because those stem cells can then differentiate into other types of cell, the tumour can continue to grow. This is not the first time that stem cells have been found in a cancer. John Dick of the Toronto General Research Institute, for instance, showed that such cells seem to cause at least some sorts of leukaemia. More recently, several researchers have found that certain brain tumours contain small subpopulations of slowly dividing cells. Per Sakariassen and his colleagues at the University of Bergen reported to the annual meeting of the American Association for Cancer Research that these cells can regenerate a tumour when injected into rats. Therefore, at least three types of cancer – breast, blood and brain – apparently rely on stem cells for their formation and growth (*The Economist*, 2004c).

Cancer stem cells, like normal stem cells, are able to pump toxins, including chemotherapy drugs, out of their internal spaces. So, even when the bulk of a tumour, made up of the fast-growing cells, is killed during chemotherapy, the stem cells escape and are able to form a new tumour. One way to kill stem cells is to force them to differentiate into more specialized cells. Such treatments already exist, but they have not been thought about in terms of stem cells. For instance, testicular tumours can be either benign or malignant; the difference lies in whether immature cells that appear to act like stem cells are present in the tumour. If a pathologist observes such cells in a biopsy, the tumour is considered malignant. The patient is then treated with drugs that drive these cells to mature; as they do so, they lose their ability for self-renewal and the tumour is no longer malignant. The objective now is to find molecules in cancer stem cells that do not exist in normal stem cells, or that act differently in healthy and unhealthy stem cells. Once they have found those molecules, researchers could develop drugs to shut down the cancerous process (*The Economist*, 2004c).

The terms of the debate

The positions of European countries

On 27 June 2002, the European Union's Council of Ministers and Parliament adopted the Sixth Framework Programme for research and development, which allocated €17.5 billion to research, including €300 million for genetics, between 1 January 2003 and 31 December 2006. From this it was anticipated to fund studies on supernumerary embryos less than two weeks old in the member states where this kind of research is authorized. In July 2002, Germany, Italy, Austria and Ireland expressed their opposition to such funding by the European Commission and imposed a moratorium until the end of 2003. At that time, they hoped that more precision would be given to the criteria determining the funding of this kind of research (Rivais, 2003). In France, the 1994 bioethics laws were to be revised by the National Assembly by mid-2004; until then, any kind of research on human embryos and embryonic stem cells was prohibited (Ferenczi, 2003).

In July 2003, after a long consultation period, the European Commission made a series of proposals that aimed at setting rigorous criteria for funding this kind of research. On 22 September 2003, the European ministers of scientific research discussed the issue but did not reach an agreement; the debate showed once again the rift between those countries that supported this kind of research, e.g. the United Kingdom and Sweden, and those that opposed it strongly (Ferenczi, 2003).

The Commission reiterated that it did not intend to suggest policy direction to member states, which are fully responsible in this area, but wanted only to define the conditions of Community funding. Research on embryonic stem cells would be financed only if it responded to major objectives and if there were no other alternative, e.g. the use of adult stem cells. Only stem cells originating from supernumerary 5–7-day-old embryos derived from *in vitro* fertilization would be involved, and the donors' consent would have to be obtained. Therefore there was no question of creating embryos through therapeutic cloning. In fact, European researchers would have to use the stock of embryos frozen and stored before 27 June 2002, the date of approval of the Sixth Framework Programme. This restriction was challenged by the countries that supported research on embryonic stem cells (Ferenczi, 2003).

In the United Kingdom, which in May 2004 opened the world's first stem cell bank, the Medical Research Council is coordinating the efforts of research teams working on different existing lines of stem cells, using standardized tools and procedures. The best teams from the United Kingdom, the United States, Australia, Canada, Israel, Singapore, the Netherlands, Finland and Sweden, as well as Germany, are associated within an international club, which aims at defining international norms. France, owing to its legislation prohibiting any research on human embryonic stem cells, is an observer within that club. With regard to Germany, this country has been opposed since 1990 to any kind of research on human embryos; however, during the summer of 2003, a new research framework was set up to enable German biologists to import current embryonic stem cell lines and to carry out research on supernumerary embryos produced *in vitro* and not intended to give rise to a human being.

The director-general of the French National Institute for Health and Medical Research (INSERM), supporting a public claim by his colleagues, has denounced the position of the French government and underlined the risk to France of being outpaced by other countries in a very promising field of research (Nau, 2003a). On 10 December 2003, within the framework of the revised law on bioethics, French parliamentarians agreed to the creation of a Biomedicine Agency, which will replace the French Organization for Grafts as well as the National Commission for Medicine and the Biology of Reproduction and Prenatal Diagnosis. In addition, they agreed on the prohibition of reproductive cloning, i.e. "any intervention aimed at giving birth to a child genetically identical to another person that is alive or has died" (Benkimoun, 2003, p. 9). It was also expected that INSERM specialists would be able to participate fully in stem cell research after the new law on bioethics was voted in by the parliament; a committee of high-level and well-known experts could review researchers' requests and authorize them to carry out their work (Nau, 2003a).

On 8–9 July 2004, the French parliament voted on the new law on bioethics, five years behind the schedule agreed on in 1994. Therapeutic cloning remains prohibited and research on human embryos remains strictly controlled. The Biomedicine Agency was to be set up on 1 January 2005. Lines of stem cells could be imported and the relevant research work would begin in the autumn of 2004. Later on, scientists would be able to work on stem cells derived from frozen human embryos that are not to be used for reproduction. The number of frozen embryos was estimated at between 100,000 and 200,000, and new measures are intended to establish the true figure and to strictly regulate the conditions of freezing human embryos (Nau and Roger, 2004).

The proposals by the European Commissioner for Science to carry out research on embryonic stem cells were submitted to the European Parliament for simple consultation in the autumn of 2003. The Parliament supported the proposals against the recommendation of its rapporteur, who suggested very tough conditions that in fact prohibited any kind of research on embryonic stem cells. However, at the Council of Ministers – the only body that can make a decision – there was no majority in favour of the European Commissioner's proposals. Italy, which was chairing the European Union until the end of 2003, proposed another text, but the Commission refused to support it. To be adopted, all 15 member states would have had to approve it; eight countries (France, the United Kingdom, Belgium, the Netherlands, Finland, Sweden, Denmark and Greece) were opposed, five (Italy, Germany, Spain, Portugal and Luxembourg) were in favour, and two (Austria and Ireland) wished to abstain (Rivais, 2003).

Ireland, which chaired the European Union for the first half of 2004, did not wish to take up this issue, owing to the passionate debate in the country. By contrast, the European Commission wanted to implement the activities indicated in the Sixth Framework Programme, adopted in June 2002, with a relevant application text. By early 2004, the Commission was to seek bids aimed at funding research on Alzheimer's and Parkinson's diseases, diabetes and cancer, which may involve the use of embryonic stem cells in the countries where this use is authorized. The proposals sent by the researchers will be submitted to a group of experts representing the 15 member states. If the experts reject the proposals, the ministers will have to make the final decision; if the required majority is not met, as probably foreseen, the Commission will take over. At the level of the Commission there was general confidence that the moratorium imposed on embryonic stem cell research in 2003 by some member states for ethical reasons would be lifted in 2004 (Rivais, 2003). As noted by Octavi Quintana, director of health research at the European Commission, progress in understanding stem cell potential will be all it takes to galvanize support for embryonic stem cell research around the globe. "The day the first clinical trials show therapeutic benefits for a patient," he stated, "opposition will completely disappear" (Kalb, 2004).

The positions of the United States and other countries

On 9 April 2001, the US President decided that research conducted on embryonic stem cells with federal funding could deal only with existing cell lines derived from embryos produced through *in vitro* fertilization. On 28 April 2004, over 200 US congresspersons, Republicans and Democrats, requested the US President to make the US regulation on the use of stem cells less stringent. Congresspeople feared that US biologists would leave the United States to work in countries with less exacting legislation. They also highlighted that around 400,000 human embryos are frozen in the United States and could therefore be used as potential sources of new stem cell lines (Nau, 2004e). By mid-June 2004, a spokesman for the US President said that restrictions on embryonic research would not be relaxed; in contrast, Massachusetts Senator John Kerry, the Democratic presidential candidate, announced that he would overturn the current policy if elected in November 2004 (Kalb, 2004).

In Australia, the first Stem Cell Summit in 2003 was held amidst a raging debate on the ethical and legal framework that should underpin stem cell research in the country. With federal legislation passed, supporting regulatory mechanisms under discussion and a strengthening commercial sector, the 2003 Summit saw many parts of the system in place for Australia to remain at the forefront of the global stem cell research endeavour (Bennett, 2003).

Polls carried out by Research Australia and Biotechnology Australia give some indication of Australians' views. Research Australia is a not-for-profit organization, independent of government, whose activities are supported by members and donors from leading research organizations, academic institutions, philanthropy, community special interest groups, peak industry bodies, biotechnology and pharmaceutical companies, small businesses and corporate Australia. Research Australia is committed to increasing grassroots awareness of the importance and benefits of health and medical research (Bennett, 2003). It found that most people have a moral objection to human cloning (82 per cent), but support stem cell research applied to disease prevention and treatment. The use of adult stem cells has strong support (70 per cent), and support for using human stem cells derived from embryos has a slight majority (53 per cent) (Bennett, 2003). Australians felt that there is not enough information available to help understand human stem cell issues better. Approximately two-thirds of the Australians interviewed responded that they were not aware or were unsure of the various issues relating to human stem cell research, and they seemed to be less aware or more uncertain about adult stem cell than about embryo stem cell research. Research Australia's on-line poll results revealed that 60 per cent of respondents did not believe that they had enough information about medical research using stem cells; only 29 per cent believed that sufficient information was available (Bennett, 2003).

The fact that many Australians are less aware of adult stem cell research may reflect the fact that many Australians gain their information through the debate presented in the media. The media's coverage largely centred on the issues around embryo stem cell research because this was the focus of ethical and community concerns. It is critical that research progress be realistically presented and not be overstated – the hype and headlines with no outcomes can lose the public's confidence and possibly impact on the research sector's credibility (Bennett, 2003).

Although Australians strongly support government regulation, only one-third felt that the use of human gene technology could be effectively regulated; most people either felt that it was not possible (49 per cent) or did not know (18 per cent). Comparison with Research America's polling on this issue shows a fairly consistent view in Australia and the United States that stem cell research to fight disease generally has majority community support. There is strong disapproval of human reproductive cloning and a strong desire for regulatory control (Bennett, 2003).

Regarding the provision of information to the community, polls showed that it should be easily accessible, realistic, accurate, clear and timely. It is also important for regulatory processes to be transparent and well communicated. This needs cooperation between research organizations, academia, private industry, government and agencies such as Biotechnology Australia, consumer groups, the media and broad-based organizations such as Research Australia (Bennett, 2003).

However, not everyone's moral code is shaped by Judaeo-Christian ethics and by the Kantian approach to human dignity (*The Economist*, 2003a). Singapore is actively recruiting people who want to work on the human aspects of biotechnology, and China, too, is said to be interested.

China's move into biotechnology has been accompanied by the introduction of biosafety regulations and a modern bioethical framework. This has been stressed by foreign observers such as Ole Doering, a bioethicist at Bochum University in Germany. In 1998, the Chinese government issued a declaration explicitly banning reproductive cloning. In addition, research on the human genome is governed by strict rules on sample collection and informed consent (*The Economist*, 2003a).

The Chinese government aims to establish national guidelines governing stem cell research. Two proposals have been made, both adapted from the British regulations, allowing therapeutic cloning. Enforcement and monitoring will also need improvement. It is true that the collaboration between Chinese scientists (e.g. Peking University Stem Cell Research Centre) and foreign research groups obliges the Chinese to abide by international rules. This is also true for publishing in international journals and attracting overseas investments (*The Economist*, 2003a).

Therapeutic and reproductive cloning

On 12 February 2004, 14 biologists working in different scientific research institutions in South Korea and led by Woo Suk Hwang and Shin Yong Moon from Seoul National University announced on the website of the periodical *Science* that they had produced the first human cloned embryo, as well as stem cells from this embryo that are capable of differentiating into somatic cells. This is a genuine advance. The South Korean researchers took 242 ovocytes from 16 women, removed the nuclei from each ovocyte and replaced them with the nuclei of cells that naturally surround the ovocytes (cell cumulus). Out of these 242 ovocytes, they obtained 200 cloned embryos, of which 30 were grown to the blastocyst stage (each more than 100 cells strong). Stem cells were identified in only 20 of these embryos and stem cell lines were isolated from a single embryo, giving rise to the body's three basic cell types (endoderm, mesoderm and ectoderm). The researchers were therefore able to extract these cells from one blastocyst and grow tissues containing all three cell types. One has to emphasize that the success rate was lower than that achieved by some teams working on human embryos resulting from *in vitro* fertilization (Nau, 2004b).

This was not the first time cloned human embryos had been produced. In November 2001, the Massachusetts-based biotechnology firm Advanced Cell Technology made several; they all died quickly. With regard to animal cloning, in November 1993 US scientists split embryos to create genetically identical twins, grew them to the 32-cell stage and then destroyed them. In July 1996, Dolly the sheep was born, the first mammal successfully cloned from adult, rather than embryonic, cells. In July 1998, University of Hawaii scientists cloned three generations of healthy mice from the nuclei of adult donor cells. In December 2001, Texas A&M University scientists created the first cloned pet, a calico kitten named CC, for Copy Cat (Lemonick, 2004).

The Korean achievement was more than a little noteworthy for two main reasons. The first is simply that their embryos did not die, whereas many experts are convinced that human clones would be very fragile. The second is that embryonic stem cells were extracted from the blastocysts and some of them were coaxed into a self-perpetuating colony. As Hwang said during a press conference at a meeting of the American Association for the Advancement of Science in Seattle in February 2004, the goal of the experiment "is not to clone humans, but to understand the causes of diseases" (Lemonick, 2004).

Several factors helped the Koreans succeed where others had failed. To start with, they had a large supply of ovocytes. The researchers lined up 16 female volunteers who found the project through its website. To avoid any taint of coercion, the women were fully informed about the research and its risks and given several opportunities to change their mind. In addition, the experiment had been approved by the national institution in charge of ethics and research on humans. The women did not receive any financial reward. In the end, the 16 women provided 242 ovocytes – many more than in any previous attempt. With such a quantity of eggs, the Korean scientists were able to test different techniques for the transfer of mature nuclei from the cumulous cells that surround ovocytes during development into enucleated ovocytes. They were therefore able to identify which technique worked best – varying the time between inserting the new nucleus and zapping it with electricity to trigger cell division, for instance, or testing different growth media (Lemonick, 2004).

Two other factors contributed to the success of the experiment. Whereas most researchers suck out an ovocyte's nucleus with a tiny pipette, Moon and Hwang made a pinhole in the cell wall and used a tiny glass needle to apply pressure and squeeze the nucleus out. The technique is more gentle on the ovocyte, allows the removal of only the DNA and leaves some of the major components of the ovocyte still inside. The second factor was that they were able to transfer a nucleus in less than a minute, which is a much better time than that achieved by most laboratories and is less likely to allow deterioration (Lemonick, 2004).

The South Korean team's scientific success will undoubtedly sharpen the debate between those who see therapeutic cloning as a potential force for good and those who see it as a step on the road to a cloned human being. Although the route from a blastocyst to a baby is a long and complex one, the South Korean breakthrough makes it more urgent than ever that legislation is passed differentiating clearly between therapeutic and reproductive cloning – permitting the former and prohibiting the latter.

With respect to the US researchers, they may be worried by the fact that the South Korean success highlighted the constraints of the American approach to regulating research with human embryonic cells derived from supernumerary embryos existing before 9 April 2001. "By this policy we are ceding leadership in what may be one of the most important medical advances of the next 10 to 15 years," stated Irving Weissman, director of Stanford's Institute for Cancer/Stem Cell Biology and Medicine. This opinion contrasts with that of Leon Kass, chairman of the US President's Council on Bioethics, who in 2002 was supported by a majority of the Council in proposing a four-year moratorium on therapeutic cloning. Many other ethicists feel strongly that medical progress is not an absolute good that can override all other values, such as the natural limits on human life and the cycle of generations (Wade, 2004b).

One should, however, remember that the first test-tube baby, born in 1978, generated an outcry about the ethics of the technique involved. Yet its great contribution to infertile families – specialist clinics in the United States alone had created more than 100,000 babies up to 2004 – outweighed the criticism. The ability to clone human embryos could follow a similar path, if it produces similar benefits. If scientists can show that therapeutic cloning saves lives, they will doubtless be able to quell the ethicists' doubts. The South Korean experiment shows that the United States is no longer the only player and it could lose its leadership (Wade, 2004b).

Scientists are indeed still learning how to coax stem cells into becoming particular types of tissue, and for many diseases they do not even know what kinds of cell they need to end up with. But some researchers underline a more immediate benefit of stem cells than their tissue-replacement ability: if they are cloned from someone with a genetic disorder, one could perform all kinds of experiments zeroing in on the DNA that is causing the disorder. According to Weissman, "this would be a transforming technology as important as recombinant DNA" (Lemonick, 2004).

At the moment, cloning mammals is a hazardous operation. It usually requires several hundred attempts to obtain a clone, and the resulting animal is often unhealthy. On 23 December 2003, US researchers at the University of Texas reported the cloning of a deer, named Dewey, which was the genetic copy of a male white-tail deer living in the south of Texas. Since the cloning of Dolly the sheep in 1996, domesticated animals (cattle, sheep, dogs and pigs) have been cloned, as have mice and rabbits. In August 2003, Italian researchers reported the first cloning of a horse and, in September of the same year, a team led by Jean-Paul Renard of the French National Institute for Agricultural Research announced the successful cloning of laboratory rats. Researchers from Advanced Cell Technology have succeeded in cloning the gaur, a wild ox being threatened with extinction in India and Malaysia. The first stage of this achievement was the culture of a skin cell from a dead animal. Then, the nuclei of several hundreds of skin cells grown in culture were transferred into enucleated ovocytes of this bovine species. One of the resulting embryos was introduced into a cow uterus and was able to develop normally. The first cloned gaur, born in November 2000 on an Iowa farm, was named Noah. In the United States, the issue of consuming cloned animals is being debated. Research nevertheless is continuing, although there does not seem to be much of a market. However, drug-producing goats and spider-silk-producing silkworms are valuable, and serious work is being done on improving the technology.

Conclusions

With respect to human ethics, the development of research on adult stem cells, spurred by therapeutic aims, may raise more formidable problems than those thought to be eliminated by forbidding research on embryonic cells. One would be dealing not with prohibiting a destruction but with the *de novo* construction of an embryonic life that shortcuts the first developmental stages of an embryo. The current ethical argument against the creation of embryos through the transfer of nuclei would become much less relevant.

The very fact of this alternative underlines the magnitude of current changes in the biological data relating to the first moments of life. Even on the definition of an embryo, science seems to hesitate. Consequently, ethical reflection and debates cannot be based just on the representation of life, which constant discoveries put in question. At the same time, people want therapies without ethical dilemmas, an absence of risk without questioning our representations of life (Renard and Bonniot de Ruisselet, 2003).

To get away from this ambiguity, one needs to distinguish the challenges of knowledge from their applications, and at the same time scrutinize the relationship between them in order to favour a regulated appropriation of innovations by society. That is why research on human therapeutic cloning may not immediately be a priority. Basic discoveries will occur far ahead of the anticipated therapeutic applications. In the absence of a clear distinction between the challenges of knowledge and economic interest, between power and the public health interest, researchers may be tempted to adopt a wait-and-see attitude (Renard and Bonniot de Ruisselet, 2003). But to forbid cloning *a priori* for a determined number of years is not an adequate approach (Renard and Bonniot de Ruisselet, 2003). Even with sensible laws, there is always a chance that cloning technology might be misused. William Gibbons, professor of obstetrics and gynaecology at Eastern Virginia Medical School, states that many useful technologies are abused every day – including automobiles and antibiotics. The solution is to legislate against the misuse, not against the technology (Lemonick, 2004).

Lessons have to be drawn from the laborious revision of bioethics law (as occurred in France in 2003), in order to find other ways of tuning the legislative process to the advancement of research. To control the continuous gradient from the creation of knowledge to innovation, new procedures of consultation must be designed involving more active participation of academic mediators, associations and trade unions between the scientific community and politicians. The inevitable conclusion is that the accumulation of biological knowledge and its very rapid evolution require a rethink of the apportionment of responsibility among researchers, politicians and citizens (Renard and Bonniot de Ruisselet, 2003).

Gene therapy

Gene therapy, which aims to cure illnesses such as cystic fibrosis, is in fact a type of genetic modification, although admittedly one that is not passed from parent to offspring. It generally meets with high social acceptance, especially when there is no cure and it is the last recourse.

For instance, a treatment for Parkinson's disease via gene therapy was tested for the first time on humans on 18 August 2003. The experiment was carried out on a 55-year-old patient at the New York Presbyterian Hospital with the approval of the US Food and Drug Administration. It was pursued in the following weeks on 11 other patients. It consists of an injection into the brain of a virus carrying the gene for the synthesis of dopamine (whose lack is the cause of the disease). The scientific community was divided about this gene therapy approach, which some neurologists considered highly risky.

The defective gene causing Huntington's disease – an affliction of the central nervous system that can cause involuntary movements and difficulty in eating – was discovered in 1993. Half a million patients worldwide are still awaiting a cure for this neurodegenerative disease. But scientists are working on a treatment that alleviates the symptoms of the disease and slows its progress. Researchers from the Center for Neurosciences in Portugal and the Swiss Federal Institute of Technology used laboratory mice infected with a rodent disease similar to Huntington's. They identified a protein that stimulates the growth of brain cells, which would counteract the damage caused by the illness. The hard part of the research was to transfer the protein into the brain; to that end, the researchers engineered a virus that, when injected into the bloodstream, "infected" the brain cells and endowed them with the ability to synthesize the neuron-growth-stimulating protein. The treatment alleviated the mice's physical symptoms and halted the progress of the disease. The scientists were also trying to use the same technique on Parkinson's patients, although effective treatments may take another five years (Witchalls, 2004).

In November 2000, Bayer AG announced a US$60 million collaboration with the US biotechnology firm Avigen, Inc. on a gene therapy treatment for haemophilia B (a deficiency in the blood-clotting protein called factor IX). The transaction, which included the purchase by Bayer of a 2.5 per cent stake in Avigen, was the latest step in an expensive revival of research at the German pharmaceutical group. It was considered a risky gamble (*Wall Street Journal Europe*, 17–18 November 2000, pp. 1 and 6). However, Bayer officials estimated that there might be just 8,000 candidates for treatment of haemophilia B with Avigen gene therapy in the United States, Europe and Japan combined. Laboratory tests of the therapy, called coagulin-B, in mice and dogs suggested that a single round of injections could restore lifelong production of factor IX at concentrations sufficient to transform severe cases of haemophilia B into moderate ones. In 2000, genetically engineered or plasma-derived versions of factor IX generated annual global sales of roughly US$350 million (*Wall Street Journal Europe*, 17–18 November 2000, pp. 1 and 6).

Bayer is already one of the world's biggest producers of factor VIII, the clotting protein used to treat haemophilia A, the commonest form of the bleeding disorder. A handful of US biotechnology companies, from Chiron Corp. and Targeted Genetics to Cell Genesys, Inc., were racing to develop factor VIII gene therapies. Bayer made a big push in this race in the mid-1990s, with a different US biotechnology ally – which has since been acquired by Chiron – but the results were really disappointing. Although factor IX gene therapy represents a far smaller market, it is also less competitive. Bayer officials believed Avigen had a lead of at least two years over potential rivals. Avigen had treated seven patients with coagulin-B in early clinical tests to demonstrate its safety and to determine a dose that would assure the sustained functioning of the replaced factor IX gene. After these tests, Bayer intended to take over worldwide development (*Wall Street Journal Europe*, 17–18 November 2000, pp. 1 and 6).

The pivotal phase-3 round of clinical trials was expected to begin by 2000, followed by applications for regulatory approval in late 2005 and a market debut the following year. Bayer officials assumed the regulators would require an unusually long follow-up period to assess the efficacy and safety of early gene therapies seeking clearance. The recruitment of patients will also be slow for such a rare disease. But if all goes according to schedule, coagulin-B may be one of the first gene therapies on the market (*Wall Street Journal Europe*, 17–18 November 2000, pp. 1 and 6).

Some recent experiments in gene therapy dealing with the repair of major deficiencies of the immune system were interrupted after the death of the patients. A French team working in this area was authorized in June 2004 to resume its experiments after the causes of death had been identified. On the other hand, extending gene therapy to germ cells to stop the disease being passed on is controversial, because of its eugenic approach.

To sum up, social acceptance of medical biotechnology and the related ethical issues discussed above underline that reliance should be placed not on bans on basic research but on the normal checks and balances, both legal and social, that should prevail in a democratic society. These have worked in the past and are likely to work in the future.

Testing drugs and ethical issues

As US companies increasingly test new drugs in other countries, they are faced with an ethical issue: they are struggling to decide what, if anything, they owe the patients who serve as test subjects. Some companies have chosen not to sell their drugs in the countries where they were tested; others have marketed their drugs there, but few patients in those countries can afford them. In recent years, companies have increasingly turned to Eastern and Central Europe in addition to countries such as India. Richard Leach, the American business manager of Russian Clinical Trials, a small company in St Petersburg, explained the appeal of doing business in those regions. Physicians in Russia are well trained but earn as little as a few hundred dollars a month, so they are eager for the money they can earn as clinical trial investigators. Patients in these countries are also eager to take part in clinical trials because their governments often do not pay for prescription drugs and few people can afford them (Kolata, 2004).

According to Lawrence O. Gostin, director of the Center for Law and the Public's Health at Georgetown and Johns Hopkins universities, "there is something troubling about 'parachute research', in which a company drops into a country, conducts its research and then leaves. It raises the question of what ethical obligation, if any, there might be to give back and make sure there is access to the drug after the trials are over" (Kolata, 2004, p. 7). And Carl B. Feldbaum, president of the US Biotechnology Industry Organization, stated: "This is something that the biotech industry, as it develops more and more drugs, will have to come to grips with. It is not that we are lacking compassion, but the economics are tough" (Kolata, 2004, p. 7).

The issue is especially difficult when it comes to drugs that do not save lives but can vastly improve the quality of life. Nobody knows for sure how many patients outside the United States have had to forgo such drugs when clinical trials ended, and companies do not give out patients' names, to protect their privacy. But the issue is of concern for company researchers and executives. Ethicists also say they are troubled. Companies must make business decisions about where to market their drugs, figuring out whether they can earn enough money to justify applying for approval, setting up business offices and hiring a sales force. If they decide not to market a drug in a given country, they are unlikely to provide it to patients there free of charge. To provide a drug for what medical professionals call compassionate use, companies must set up a distribution system, train doctors to administer the drugs, monitor patients for adverse effects and track the results (Kolata, 2004).

In the United States, patients participating in clinical trials often continue to receive the drug being tested until it is approved. After that, they can buy it or, if they cannot afford it, apply to the programmes that most companies offer to help people obtain the drug. However, with the exception of anti-HIV/AIDS drugs, which companies provide free or at low cost to patients in poor countries, there is no consensus on what to do internationally, especially when drugs are not life saving (Kolata, 2004).

Biopharming

Consumer advocates fear that plant-grown drugs and industrial chemicals will end up in their meals; critics have called this Pharmageddon. In October 2002, inspectors from the US Department of Agriculture (USDA) Animal and Plant Health Inspection Service (APHIS) discovered transgenic maize growing in a soybean field at a site in Nebraska. The field had been used the previous year by the biotechnology company ProdiGene, Inc. for field-testing a transgenic maize variety containing a vaccine against pig diarrhoea. APHIS instructed ProdiGene to remove the maize plants from the field, despite the fact the plants had no viable seed. However, the soybeans were harvested and taken to a storage facility before all of the maize was removed. APHIS immediately placed a hold on the soybeans so that these materials would not enter the human or animal food chains. Another breach of the US regulations was discovered at a ProdiGene test site in Iowa in September 2002, and the maize plants were removed from the field earlier in the season. The contaminated soybean batches did not enter the human or animal food supply chain.

In December 2002, ProdiGene agreed to pay USDA more than US$3 million for breaching the Plant Protection Act (approved in 2000 to regulate the production and transportation of transgenic plants). The company will pay a civil penalty of US$250,000 and will reimburse USDA for all the costs of collecting and destroying the contaminated soybeans and cleaning the storage facility and all equipment. ProdiGene also agreed to a US$1 million bond and higher compliance standards, including additional approvals before field-testing and harvesting transgenic material. The company was expected to develop a written compliance programme with USDA to ensure that its employees, agents and managers are aware of, and comply with, the Plant Protection Act, federal regulations and permit conditions. These incidents rattled the industry and fuelled the debate about the coexistence of biopharming and conventional agriculture. Some key players, among them Monsanto and Dow Agrosciences, prefer to grow their pharmamaize in isolated areas of Arizona, California and Washington State, rather than in the Corn Belt.

In August 2002, USDA created a new Biotechnology Regulatory Services Unit within APHIS for regulating and facilitating biotechnology. Draft guidance to industry on drugs, biologicals and medical devices derived from bio-engineered plants for use in humans and animals was published in September 2002. USDA also set up a new unit in the Foreign Agricultural Service to deal with biotechnology trade issues.

In November 2002, the Biotechnology Industry Organization issued a policy statement regarding plants that produce pharmaceutical and industrial products. The statement supports strong controls on the management of transgenic crops, including:

- appropriate measures for confinement or containment, including spatial isolation from major areas of crop production intended for animal or human consumption;
- encouraging alternative approaches to this issue that would deliver at least equivalent assurances for the integrity of the food supply and export markets;
- USDA/APHIS permit requirements for the inter-state movement, field-testing and commercial planting of all regulated articles that are not intended to be used for human or animal consumption;
- submission by companies of detailed confinement and handling plans and standard operating procedures with each of their mandatory permit applications to USDA/APHIS;
- such plans and procedures to be mandatory permit conditions, subject to mandatory audit and inspection;
- FDA and USDA guidance for industry on the production and use of biopharmed plants.

In March 2003, USDA proposed a set of rigid, one-size-fits-all rules. Land used to grow biopharmed maize, for instance, would have to lie fallow for the following season, a requirement that would promote soil erosion and whose expense would discourage many farmers from biopharming. Also proposed is a costly requirement of separate equipment for biopharming. These new rules will also step up inspections of biopharms and extend the buffer zone between genetically engineered maize and food crops to 1.5 km. However, opponents state that this buffer zone is not wide enough to prevent cross-pollination, and a coalition of 11 environmental groups has filed a suit against USDA. They wanted to ban the use of food crops for pharmaceutical purposes and restrict the plants to greenhouses. The chief scientist of Monsanto Protein Technologies argued that, if such measures were enforced, it would set back the industry 12 to 20 years.

The Canadian Food Inspection Agency is working with other government agencies (including Health Canada) and their counterparts in the United States, as well as with the rapidly growing molecular farming industry on both sides of the border, to develop regulations that will ensure that benefits can be enjoyed without putting the environment or human health at risk – a regulatory system that is not so restrictive as to discourage the molecular farming industry from investing the millions of dollars required to make these exciting new products a reality.

Gene escape from a biopharmed crop to a conventional one would occur only if a certain gene from the crop confers a selective advantage on the recipient – an occurrence that should be uncommon with biopharming, where most often the added gene would make the plant less fit and less able to proliferate. Gene transfer is an age-old concern of farmers who have learnt how to prevent pollen cross-contamination when necessary for commercial reasons. For instance, in order to maintain the highest level of genetic purity, distinct varieties of self-pollinated crops such as wheat, rice, soybeans and barley need to be separated by at least 60 feet.

What is the likelihood of consumers sustaining harm, even in a worst-case scenario? Several highly improbable events would have to occur. First, the active drug would have to be present in the food in sufficient amounts to exert an adverse effect via direct toxicity or allergy. Secondly, the active agent would need to survive milling and other processing, as well as cooking. Thirdly, it would need to have a biological effect when ingested.

8

The globalization of regulatory standards and ethical norms: Solidarity with developing nations

Philippe Kourilsky, director-general of the Institut Pasteur in Paris, has underlined that the well-known economic difficulties, as well as the lack of solidarity from the industrialized world, that hamper health-care policies in developing countries are compounded by the overregulation and ethical standards imposed by the rich countries. In other words, "the ethics of the North sacrifices the patients of the South" (Kourilsky, 2004). Citing the glaring gap between the health-care needs of developing countries and the scarcity of funds collected at international level to meet these needs, Kourilsky (2004) mentions that 700,000 children still die from measles and its complications every year, and an effective vaccine costs only a few cents. Thanks to a donation from the Bill and Melinda Gates Foundation (which represents about 1 per cent of the cost of the war in Iraq), the World Health Organization and UNICEF (the United Nations Children Fund) were able to fulfil their vaccination mission.

There are many reasons for the currently disastrous health situation in the poor countries. The selfishness of rich countries is obviously one reason; another one is the lack of clearly stated and efficient health-care policies in developing countries. Pharmaceutical companies and groups should not be blamed systematically. They need to make their policy of patent management more flexible, but they have to deal with market rules in a very competitive deregulated market; they need to procure medicines and vaccines at a minimum cost, but they are not responsible for solving the world's health problems. Vaccines are very much a "neglected" area, accounting for only 2 per cent of the world's pharmaceutical industry market. It is no wonder therefore that, with 20–50 times less funding than is invested in medicines, an anti-HIV/AIDS vaccine is still a hypothetical venture, whereas anti-retroviral drugs were commercialized as of 1996 (Kourilsky, 2004).

Another obstacle, generally less obvious and known to the public, relates to the regulatory framework set up by the US Food and Drug Administration and the European Agency for the Evaluation of Medicinal Products among the main regulatory agencies. These independent bodies set the standards that govern the research, development and manufacture of drugs and vaccines. These standards are constantly raised owing to the logic of extreme safety prevailing in the industrialized countries. Costs follow suit and increased three-fold over a period of 15 years with respect to the clinical and pharmaceutical development of a vaccine, which currently costs several hundred million dollars. These agencies have no counter-power. The safety benefits are seldom evaluated and one wonders who these ever-increasing standards are protecting: the vaccinees and the patients, or the producers and the regulatory authorities themselves (Kourilsky, 2004).

While imposing regulatory standards on the South, the North is creating protectionist barriers, because the South cannot produce at lower cost so as to export its products to the North. Moreover, unable to achieve the appropriate standards, the poor countries often refrain from producing for themselves even though they are not prevented from doing so (Kourilsky, 2004).

However, the globalization of regulatory standards and frameworks is accompanied by a globalization of ethics. Those in favour of universal ethics are opposed to those who wish to adapt ethical norms to local conditions, rejecting any hint of a "double standard". They confuse regulatory standards and ethical norms, and in their view all clinical trials, pharmaceutical development processes and manufacturing practices should be the same in the North and the South. This approach, which looks rather like an ideal goal, gives rise to dramatic problems. For instance, to combat three neglected diseases, a consortium with a budget of US$50 million (DNDi, the Institut Pasteur and Médecins sans frontières) hopes to develop eight medicines. This would be an impossible mission if they followed "western" norms, under which a drug costs about half a million dollars to develop. Another example is the withdrawal of an anti-rotavirus vaccine from the US market in 1999 because of a small number of undesirable side-effects affecting 20 out of 500,000 vaccinated children. Other vaccines are undergoing clinical trials on 80,000 volunteers; in the meantime, 500,000 children die annually because of the lack of a vaccine. What can be done? Who should determine what is a reasonable approach? It should be up to the developing countries and their citizens to make the decision, because it is mainly they who are suffering. The decision should not be based on a universal ethics – that of "western" countries (Kourilsky, 2004).

On the other hand, developing countries should not become a dumping ground, where one can do away with standards and norms of safety and ethics. There is no question of suggesting a lax approach that might lead to health disasters. The real issue is to make a distinction between regulatory standards and ethical ones. For instance, according to Kourilsky (2004), what is wrong is not using vaccines developed 40 or 50 years ago but still quite effective, even though they are not in conformity with current norms. Kourilsky is of the opinion that, without relying on international organizations or on industrial groups, one should evaluate the true safety benefits derived from the existing regulations and compare them with the costs they imply and generate. One should also review the ethical transactions between individual and collective benefits, on the basis of local issues and not of general ideas that would make the precautionary principle prevail over the evaluation of risks and benefits; the latter approach is the real ethical fraud. Finally, emphasis should be laid on solidarity and generosity at all levels, qualities that tend to retreat as wealth increases (Kourilsky, 2004).

With respect to this last point, it is worth mentioning that some drug inventors are willing to donate their royalties to help the poor. This is, for instance, the case with Professor Gordon H. Sato, a cell biologist and a member of the US National Academy of Sciences. Sato was the co-inventor of the anti-cancer drug Erbitux – a long-awaited drug from Im-Clone Systems approved in 2004. He worked in the early 1980s on this drug, which would become a major new treatment for colon cancer. The work was carried out at Sato's laboratory at the University of California, as well as by John Mendelsohn, president of the M.D. Anderson Cancer Center in Houston. Once the laboratory work was finished, Sato left San Diego and moved on to other projects. It was Mendelsohn who turned the laboratory results into a useful cancer treatment. He helped set up the first clinical trials, and he arranged to license the drug to ImClone Systems.

The royalty rate for Erbitux is believed to be 1 per cent; so, if sales reach the expected level of several hundred million dollars a year, Sato could receive several hundred thousand dollars a year. Royalty payments are set to end by mid-2007, when the patent expires, although the university will probably apply for an extension. Since Eritrea's independence in the early 1990s, Sato has spent more than half of every year there and intends to spend his royalties on the Manzanar Project, which aims to produce food in a coastal Eritrean village. Indeed, Sato has stated that he has spent about US$500,000 of his own money on this project (Pollack, 2004a).

References

Adam, D. (2003) "Brazil's R&D agenda. Under new management", *Nature* 423(6938): 367–380.

Adhikari, R. (2004) "Biotech grows – bio-manufacturing lags behind", *European Biotechnology News* 3(3): 42–44.

Adiga, A. (2004) "Market jitters. SARS was an economic disaster. Could bird flu be as bad?" *Time*, 9 February, p. 21.

Arnold, W. (2003) "Singapore builds a better scientist trap", *International Herald Tribune*, 27 August, pp. 1 and 8.

Benkimoun, P. (2003) "Les députées autorisent le recours expérimental au 'bébé-médicament'. Des cellules souches pourront être prélevées pour soigner un autre enfant", *Le Monde*, 12 December, p. 9.

Bennett, C. (2003) "Responsibility to the public: Why communicating the facts is essential to the progress of stem cell research", *Australasian Biotechnology* 13(3): 32–33.

Bhardwaj, M. (2003) "Global bioethics and international governance of biotechnology", *Asian Biotechnology and Development Review* 6(1): 39–53.

Bobin, F. (2003) "Chine: Une puissance scientifique encore immature", *Le Monde*, 17 October, p. 15.

Bonte-Friedheim, R. and Ekdahl, K. (2004) "World health. A flu threat harder to stop than SARS", *International Herald Tribune*, 12 April, p. 8.

Bowe, C. (2004) "Cholesterol drug helps Pfizer sales soar to $12.5 bn", *Financial Times* (London), 21 April, p. 15.

Boyd, T. (2003) "Novel seaweed compounds creating a Trans-Tasman link", *Australasian Biotechnology* 13(3): 39–42.

Burton, J. (2004) "Novartis decision boosts Singapore ambitions", *Financial Times*, 6 July, p. 6.

143

Choisy, M., Woelk, C. H., Guegan, J. F. and Robertson, D. (2004) "Comparative study of adaptative molecular evolution in different human immunodeficiency virus groups and subtypes", *Journal of Virology* 78(4): 1962–1970.

Clendennen, S. K., López-Gómez, R., Gómez-Lim, M., Arntzen, C. J. and May, G. D. (1998) "The abundant 31-kilodalton banana pulp protein is homologous to class-III acidic chitinases", *Phytochemistry* 47: 613–619.

Cookson, C. (2003) "Genetics. DNA profiling: A double-edged sword of justice", *Financial Times*, 19 December, p. 9.

Covello, P. (2003) "Gene discovery and the supply of plant-derived drugs", *PBI Bulletin* 1: 4–5.

Dando, G. and Devine, P. (2003) "Emerging companies and pre-seed investment", *Australasian Biotechnology* 13(3): 24–26.

Dyer, G. (2003a) "EU approval process for drugs is a 'farce'", *Financial Times*, 28 October, p. 17.

Dyer, G. (2003b) "Research boost for Viagra challenges", *Financial Times*, 17 November, p. 1.

Dyer, G. (2003c) "Aventis cancer drug has positive results", *Financial Times*, 6–7 December, p. 8.

Dyer, G. (2004) "UK loses sector leader … but Europe gains a major biotechnology player", *Financial Times*, 19 May, p. 20.

Elegant, S. (2004) "Avian flu. Gauging the threat. Is a human pandemic next?", *Time*, 9 February, pp. 14–20.

European Commission (2002) *Wonders of life. Stories from life sciences research (from the Fourth and Fifth Framework Programmes)*. Luxembourg: Office of Official Publications of the European Communities.

Facchini, P. J. (2003) "Prairie poppy fields: Why research on opium poppy is important and relevant in Canada", *PBI Bulletin* 1: 6–8.

Ferenczi, T. (2003) "Le financement de la recherche sur les cellules souches embryonnaires divise les Européens", *Le Monde*, 24 September, p. 7.

Firn, D. (2003) "European biotech's dogged fight to keep up", *Financial Times*, 19 December, p. 8.

Firn, D. and Minder, R. (2004) "UCB buys entry into biotech market. UCB liked the data so much it bought the company", *Financial Times*, 19 May, pp. 19 and 20.

Fogher, C., et al. (1999) *Abstracts International Symposium on Plant Genetic Engineering*, Havana, Cuba, 6–10 December, p. 224.

Foster, L. (2004) "Selling tight skin to the ageing masses", *Financial Times*, 21–22 February, p. 8.

Francisco, A. de (2004) "French bioindustry: Diverse and promising", *European Biotechnology News* 3(3): 38–41.

Gabrielczyk, T. (2004) "New spirit of entrepreneurship", *European Biotechnology News* 3(3): 3.

Gorman, C. and Noble, K. (2004) "Why some are getting diabetes", *Time*, 12 January, pp. 37–43.

Graff, G. D. and Newcomb, J. (2003) *Agricultural biotechnology at the crossroads. Part I. The changing structure of the industry*. Cambridge, MA: BioEconomic Research Associates (bio-era™).

Greco, A. (2003) "From bench to boardroom: Promoting Brazilian biotech", *Science* 300: 1366–1367.

Griffith, V. (2003a) "Aventis and Genta add to biotech surge", *Financial Times*, 11 September, p. 19.

Griffith, V. (2003b) "Biogen Idec looks for critical mass in its pipeline", *Financial Times*, 19 December, p. 22.

Guterl, F. (2004) "Clipping its wings. Scientists hope a new technique will help them develop a vaccine against the bird flu virus before it leaps to humans", *Newsweek*, 9 February, pp. 36–40.

Hilgenfeld, R., et al. (2003) "Coronavirus main proteinase (3CLpro) structure: Basis for design of anti-SARS drugs", *Science*, 13 June, pp. 1763–1767.

Japan Bioindustry Association (2003), *JBA Letters* 20(1).

Kahn, A. (2002) "La France veut mieux tirer parti du potentiel scientifique chinois", *Le Monde*, 13 November, p. III.

Kahn, A. (2003a) "La Chine se hisse au troisième rang mondial en recherche et développement", *Le Monde*, 4 November, p. V.

Kahn, A. (2003b) "Un plan sur cinq ans pour les biotechnologies", *Le Monde*, 28–29 September, p. 17.

Kalb, C. (2004) "The life in a cell", *Newsweek*, 28 June, pp. 50–51.

Khandelwal, A., et al. (1999) *Abstracts International Symposium on Plant Genetic Engineering*, Havana, Cuba, 6–10 December, p. 223.

Kolata, G. (2004) "The ethics of testing drugs on patients who cannot afford them", *New York Times–Le Monde*, 21–22 March, p. 7.

Kourilsky, P. (2004) "L'éthique du Nord sacrifie les malades du Sud", *Le Monde*, 8–9 February, pp. 1–14.

Langridge, W. H. R. (2000) "Edible vaccines", *Scientific American* 283(3): 48–53.

Larrick, J. W., Yu, L., Chen, J., Jaiswal, S. and Wycoff, K. (2000) "Production of antibodies in transgenic plants", *Biotecnología Aplicada* 17(1): 45–46.

Lean, G. (2004) "GM rice to be grown for medicine", *The Independent*, 1 February, p. 2.

Lemonick, M. D. (2003) "Tomato vaccine", *Time*, 25 November.

Lemonick, M. D. (2004) "Cloning gets closer. A Korean team clones human cells to fight disease – kindling fears someone will go too far", *Time*, 23 February, pp. 46–48.

Leroy, Eric M., et al. (2004) "Multiple Ebola virus transmission events and rapid decline of Central African wildlife", *Science* 303(5655).

Lorelle, V. (1999a) "Les start-up de biotechnologies seront les groupes pharmaceutiques de demain. Des molécules qui piègent la cocaïne dans le sang", *Le Monde*, 10 November, p. 20.

Lorelle, V. (1999b) "Monsanto et Pharmacia & Upjohn se marient pour séduire les marchés", *Le Monde*, 21 December, p. 21.

Lorelle, V. (2000) "Après sa fusion avec Monsanto, Pharmacia doit relever le défi des OGM. Aux Etats-Unis, les cultures transformées reculent", *Le Monde*, 6 April, p. 17.

Lorelle, V. (2003) "Les sociétés de biotechnologie s'emparent des maladies rares", *Le Monde*, 2 August, p. 10.

Lorelle, V. and Ducourtieux, C. (2003) "Les déboires financiers et boursiers des sociétés de biotechnologie", *Le Monde*, 29 March, p. 26.

Luneau, D. (2003) "Nantes mise sur les sciences de la vie. La ville accueille le septième Carrefour européen des biotechnologies", *Le Monde*, 30 September, p. 18.

Macer, D. (1998) *Bioethics is love of life*. Eubios Ethics Institute.

Mamou, Y. (2003a) "Aventis se restructure pour économiser 500 millions d'euros par an d'ici à 2006", *Le Monde*, 10 December, p. 18.

Mamou, Y. (2003b) "Le laboratoire Merck compte plus que jamais sur ses propres forces", *Le Monde*, 12 December, p. 26.

Mamou, Y. (2004a) "Avec Hespérion, Cerep accroît son offre de services aux grands laboratoires", *Le Monde*, 21 January, p. 20.

Mamou, Y. (2004b) "L'inévitable concentration de l'industrie pharmaceutique", *Le Monde*, 14 February, p. 18.

Mamou, Y. (2004c) "La Suisse en piste pour un leadership mondial", *Le Monde*, 25 March, p. 22.

Mamou, Y. (2004d) "Un répit pour les laboratoires français", *Le Monde*, 27 April, p. 19.

Mamou, Y. (2004e) "Les sociétés de biotechnologie américaines relèvent la tête. Genentech, de la start-up au puissant laboratoire. Quand chercheurs, entrepreneurs et banquiers s'associent", *Le Monde*, 26 May, p. 20.

Marcelo, R. (2003) "India beckons as a test bed for western drug companies", *Financial Times*, 14 October, p. 18.

Marti, S. (2003) "La Chine, atelier du monde, joue la carte du high-tech", *Le Monde*, 8 April, p. I.

Mason, H. S., Warzecha, H., Mor, T. and Arntzen, C. J. (2002) "Edible plant vaccines: Applications for prophylactic and therapeutic molecular medicine", *Trends in Molecular Medicine* 8(7): 324–329.

Menassa, R., et al. (1999) *Abstracts International Symposium on Plant Genetic Engineering*, Havana, Cuba, 6–10 December, p. 209.

Moreno, G. (2003) "Mexican food exporters balking at US bioterrorism regulations", *New Straits Times*, 10 October, p. B.17.

Nau, J.-Y. (2003a) "Sans loi de bioéthique, la recherche française ne sera plus compétitive", *Le Monde*, 3 October, p. 26.

Nau, J.-Y. (2003b) "Par crainte du bioterrorisme, les Etats-Unis expérimentent un vaccin contre le virus d'Ebola", *Le Monde*, 22 November, p. 26.

Nau, J.-Y. (2004a) "Un mea culpa de l'Académie des sciences chinoise", *Le Monde*, 31 January, p. 6.

Nau, J.-Y. (2004b) "Des biologistes coréens ont créé des embryons humaines par clonage et obtenu des cellules-souches", *Le Monde*, 13 February, p. 23.

Nau, J.-Y. (2004c) "Le génome du rat séquencé après ceux de l'homme et de la souris", *Le Monde*, 2 April, p. 25.

Nau, J.-Y. (2004d) "Plus de 21.000 gènes humains décrits avec précision", *Le Monde*, 21 April, p. 23.

Nau, J.-Y. (2004e) "Des cellules souches de la moelle osseuse pourraient régénérer le cerveau. Une importante découverte américaine", *Le Monde*, 7 May, p. 24.

Nau, J.-Y. and Roger, P. (2004) "Après trois années de débat, la révision de la loi sur la bioéthique est adoptée par le Parlement", *Le Monde*, 9 July, p. 8.

Oomah, B. D. (2003) "Isolation, characterization and assessment of secondary metabolites from plants for use in human health", *PBI Bulletin* 1: 26–27.

Orange, M. (2004) "Sanofi-Aventis devient le troisième groupe pharmaceutique mondial", *Le Monde*, 27 April, p. 19.

Peña, A. (2004) "ADN: Un tesoro de ultratumba", *La Gaceta de los Negocios*, 19 May, p. 53.

Pilling, D. (2004) "Pharmas seek new dawn in Land of the Rising Sun. The Yamanouchi–Fujisawa merger is a sign Japan's drug policies are relaxing", *Financial Times*, 25 February, p. 21.

Pollack, A. (2004a) "Profits from cancer drug may help feed the hungry", *New York Times–Le Monde*, 21–22 March, p. 7.

Pollack, A. (2004b) "Cuban cancer drugs to be licensed in U.S.", *International Herald Tribune*, 16 July, p. 15.

Pons, P. (2000) "Le groupe pharmaceutique Daiichi et Fujitsu s'allient dans la recherche sur le génome. Le Japon veut rattraper son retard sur les Etats-Unis", *Le Monde*, 1 November, p. 19.

Potter, V. R. (1970) "Bioethics: The science of survival", *Perspectives in Biology and Medicine* 14(1): 127–153.

Pouletty, P. (2004) "France boosts its attractiveness", *European Biotechnology News* 3(3): 34–38.

Pujol Gebelli, X. (2003) "Yondelis se acabará aprobando tarde o temprano. José María Fernández Sousa-Faro, presidente de PharmaMar. Dinero a contracorriente bajo el mar", *Boletín SEBBM* (Sociedad Española de Bioquímica y Biología Molecular), 137: 20–22.

Renard, J.-P. and Bonniot de Ruisselet, J. (2003) "L'embryon humain face à la recherche", *Le Monde*, 8 July, p. 14.

Rivais, R. (2003) "Bruxelles entend financer les recherches sur les cellules-souches embryonnaires humaines", *Le Monde*, 21–22 December, p. 6.

Roosevelt, M. (2003) "Cures on the cob", *Time*, 26 May, pp. 56–57.

Swann, C. (2004) "Lunch with the FT/James Watson. The man who would have us play God", *Financial Times*, 31 January–1 February, p. W3.

Tait, N. (2000) "Pharmacia lifted by anti-arthritis drug sales", *Financial Times*, 31 October, p. 22.

The Economist (2002a) "Chinese biotechnology. Biotech's yin and yang", *The Economist*, 14 December, pp. 75–77.

The Economist (2002b) "A drug of one's own", *The Economist*, 14 December, pp. 35–36.

The Economist (2003a) "Climbing the helical staircase. A survey of biotechnology", *The Economist*, 29 March, pp. 3–18.

The Economist (2003b) "Face value. Food, drugs and economics", *The Economist*, 23 August, p. 50.

The Economist (2003c) "Biotechnology. Carbon copy: Making generic biotech drugs will be a tough business", *The Economist*, 11 October, pp. 70–71.

The Economist (2003d) "Hepatitis C. Needles and haystacks", *The Economist*, 1 November, pp. 79–80.

The Economist (2003e) "Biotech in Cuba. Truly revolutionary", *The Economist*, 29 November, pp. 70–71.

The Economist (2004a) "Special report. Venture capital. After the drought", *The Economist*, 3 April, pp. 65–67.

The Economist (2004b) "Diabetes. Headed off at the pass?", *The Economist*, 3 April, p. 79.

The Economist (2004c) "Cancer research. Stemming the tide", *The Economist*, 10 April, p. 73.

Wade, N. (2004a) "As SARS evolved, its potency soared. Attack mechanism startles scientists", *International Herald Tribune*, 31 January–1 February, pp. 1 and 4.

Wade, N. (2004b) "Unease over cloning may cost U.S. influence", *International Herald Tribune*, 17 February, p. 4.

Walmsley, A. M. and Arntzen, C. J. (2003) "Plant cell factories and mucosal vaccines", *Current Opinion in Biotechnology* 14(2): 145–150.

Witchalls, C. (2004) "Friendly brain bug. A virus halts the progress of Huntington's in mice", *Newsweek*, 12 January, p. 43.

Index